Science & Math EVENTS:

CONNECTING & COMPETING

National Science Teachers Association

Center for Teaching
The Westminster Schools

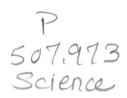

This book has been produced by the National Science Teachers Association, 1742 Connecticut Avenue, N.W., Washington, DC 20009. Assistant Executive Director for Publications Phyllis R. Marcuccio; Managing Editor of Special Publications Shirley Watt Ireton; Free-lance Editors Deborah C. Fort, Pat Tschirhart-Spangler, Michael Byrnes; Designer Bob McMeans.

This material is based upon work supported by the National Science Foundation under Grant No. TE-8751313. Any opinions, findings, and conclusions or recommendations expressed in this material are those of the authors and do not necessarily reflect the views of the National Science Foundation.

Library of Congress Catalog Card Number 90-060246

ISBN Number 0-87355-090-0

NSTA 7.00
10/1/92

TABLE OF CONTENTS

Science Competitions—Nationwide and Worldwide

FOREWORD

You hold in your hands a book that lists and analyzes opportunities for teenagers to participate in organized science and math activities within and without the classroom. Students may participate at the local or school level by joining clubs and interest groups and by entering fairs; at the national level, by entering competitions of various sorts; and at the international level, by competing in olympiads and other kinds of contests.

Science is often collaborative. For example, C. Robert Ballard emphasizes that it was *"teamwork"* that devised the "leading edge" technology that led to the robotic underwater explorer Jason, the submarine that found the *Titanic*, ancient Roman shipwrecks in the Mediterranean, and the German battleship *Bismarck*. Hence, our subtitle: *Connecting* and Competing. This cooperation, which numerous observers have noted among mature scientists, on occasion is—or should be—part of science and math events for students as well. On average, the production of one scientific breakthrough requires the work of a team of eight individuals—individuals with different skills, each making a unique contribution (Derek de Solla Price, 1961/1975). Well-run science clubs can bring peers together for invaluable support and feedback—as did the biology, chemistry, math, and premed clubs at Stuyvesant High School in New York City for student Joshua Lederberg, later Nobelist and Medal of Science winner. Likewise, a soundly organized science fair can make the research and creation of the exhibit a learning experience, which leads to a deeper interest in the subject tackled.

On the other hand, as readers of James D. Watson's *The Double Helix* (1968) can testify, science is also, at times, competitive. Though at a given time, scientists may be wrestling with similar problems, they may be doing so absolutely independently, with diverging equipment and approaches, on different continents, without collaborating—at least until publication of results. Or another situation may apply, similar to that in organized sports, where a group of athletes play as a team, with victory as their common goal, but sometimes, often, winning through the efforts of outstanding individuals. By playing the game (or entering a science competition), contestants may gain in many ways. Competitors may improve their skills not only by trying them out but also by receiving expert coaching.

Thus, the potential for significant education exists in science and mathematics activities of many sorts. The best of these activities offer experiences that lead to learning by doing and through individual efforts, often with the help of teachers, other mentors, peers, or a combination of all three. While this volume offers both help as a guide to science opportunities that can help students grow and caution against foolishly organized activities and contests, we make no pretense at completeness.

Interested readers should also consult other sources of information about science and math programs—both formally and informally organized—for students in middle school through high school. See the box accompanying Gene Kutscher's paper on science research projects (page 67) for information on ways to find and take advantage of other science and math opportunities, particularly those occurring at times other than in school hours during the academic year.

PART I—INSPIRATION

The opening section of *Science and Math Events: Connecting and Competing* presents the views of four experienced high school teachers on what inspires students to join science fairs and to participate in competitions at the school, city, state, national, and international levels. Leonard M. Krause analyzes the value of feedback from student to student that occurs when science interest groups or clubs come together; John A. Blakeman offers the argument that science presentations and competitions, properly run, fulfill a function similar to that of the scientist's publication of results. Nicholas D. Frankovits, mentor to numerous winners of local, regional, national, and international contests, speaks of the skills and confidence that contestants gain through science competitions. Four of Frankovits' students testify about their experiences in competitions. And Kutscher, mentor to many students in science competitions at many levels, writes of the kind of learning and scholarship that is fostered by preparing to compete—whether the student eventually chooses to enter or not.

PART II—INVOLVEMENT

The lion's share of the book covers the nuts and bolts of involvement. Here, readers will find information about participation in science interest groups and clubs. Blakeman writes of the former; Peter G. Bruecken of the latter. We also include a sample science club constitution from NSTA's useful but long out-of-print *Sponsor's Guidebook*

(1971, from the Future Scientists of America Club) and an analysis of the results of a recent survey on science clubs conducted by *NSTA Reports!* Next, Mark A. Wagner describes out-of-school science clubs.

Then, Nancy C. Aiello and Brian E. Hansen offer advice on the value of a successful science fair and the procedures that help to get one started. Their piece is accompanied by supplementary materials— a judging form and a student science fair application. Blakeman's observations about poster projects and science symposia follow. Then, Kutscher returns to explain the procedures of science research projects, and finally, Donald R. Daugs writes about "invention fairs."

In the third section of part II, individual entries describe 28 science and 4 math competitions. Listed here are 21 science and 3 math contests held nationally, and 7 science and 1 math competitions, internationally. Introducing this section is Arthur Eisenkraft, science teacher and frequent adviser to students competing in U.S. contests like Duracell NSTA and worldwide ones like the Physics Olympiad. Participation at all levels can extend the depth and breadth of students' interest in and commitment to scientific investigations. It can also lead to national competitions, such as those sponsored by DuPont, Westinghouse, and many other organizations, and eventually to international contests.

To gather information for the 32 individual entries, NSTA circulated surveys to 32 contest organizers, 321 student participants, 272 of their mentor/teachers, and 67 judges. Each entry contains information on the competition's background and funding, those eligible to compete, its rules, deadlines, and fees, its awards, and its special features. Contests not only offer learning experiences and acclaim for participants; competition can also lead to substantial awards, from cash to college scholarships to many other prizes that range from medals and certificates to expenses-paid trips.

In all cases, contest organizers checked drafts of their entries for accuracy and completeness, but details change from year to year. *So, prospective contestants must still write directly to the organizers for details and rules—and most importantly, deadlines—specific to particular years.*

Surveys returned from participants revealed interesting concerns and complexities. (Completed examples of all surveys sent are available in the appendixes.) For example, some judges' jobs were as challenging as Nobelist Glenn T. Seaborg's, who was one of the evaluators of the Westinghouse Science Search and who looked especially for "basic understanding and originality," while others' tasks were as straightforward as looking over multiple-choice responses or checking calculations submitted by Olympiad teams. Twenty-two judges told NSTA what special qualities they sought, whether they gave credit for effort, whether they saw pain for the "losers," what method of presentation they preferred—for example, anonymous or by interview.

NSTA heard also from 71 teachers and/or mentors, although many, many more contributed without official recognition. Advisors revealed how much help they think is appropriate and how much time they allot to assisting students entered in national and international contests. They also told us how they encouraged students to enter contests and to stick with their commitments, and whether they thought the experience of losing was disheartening. Some also addressed the issue of student safety—one often overlooked but absolutely basic. Without the generosity of these professionals and volunteers, few students would come to know the joy of science clubs, of participating in science fairs, or competing at the city, state, national or international level.

But our best information may have come from the students themselves. Among other things, 108 respondents told NSTA where they got ideas for their projects, how much time they spent, where they worked, and whether they hoped—eventually—to become scientists, mathematicians, or teachers of science and/or math. Examples of all surveys appear in the appendices.

PART III—ADVICE FROM HINDSIGHT

In this section, you will find the results of one final set of questionnaires. NSTA polled 164 U.S. Nobel and Medal of Science winners about their experiences with organized science and math enterprises beyond the regular curriculum. Those 89 answering—more than half of the group surveyed—revealed a striking disharmony on whether or not they favored science competitions as a means of inspiring students. Half were undecided; 15 percent disapproved; 35 percent approved. The large majority of our respondents never participated personally, however, because science fairs and contests (though not clubs and interest groups) are relatively new to American science education.

ACKNOWLEDGMENTS

We deeply appreciate the help of many teachers, associations, and others. Of particular help were Doug Hunt and Linda Durham of the National Association of Secondary School Principals, who sent annotated lists of approved competitions and who helped track elusive contests; Joe Dasbach, of the American Association for the

Advancement of Science and Julian Stanley, of the Study of Mathematically Precocious Youth at Johns Hopkins University, whose early guidance set us on the right path; Carol Luszcz and E. G. Sherburne, Jr., of Science Service, without whose help at all stages the project would have been much less accurate.

Editorial and clerical staff at NSTA also worked hard and carefully to make this volume complete. We thank Pat Tschirhart-Spangler, Sovita Chander, and Michael Byrnes for their work on the national competitions and Deborah C. Fort for her efforts on the international science events, science club surveys, and "Advice from Hindsight." Thanks go to NSTA Executive Director Bill G. Aldridge, Assistant Executive Director for Publications Phyllis R. Marcuccio, and Managing Editor of Special Publications Shirley Watt Ireton.

A special note of gratitude goes to the National Science Foundation, whose funding made possible both our research and this publication, which results from a 1988 teacher's summer conference held at Long Island University under the direction of Edward D. Lozansky.

NSTA Special Publications Staff
Spring, 1990

REFERENCE

De Solla Price, Derek. (1961/1975). *Science since Babylon* (enlarged ed.). New Haven: Yale University Press.

PART I:

Inspiration:
Why Join In?

JOIN—TO GET FEEDBACK FROM ONE'S PEERS

Leonard M. Krause

Science clubs, symposia, conferences, and other forms of activities for youth have been part of the education scene for many decades. What is it that attracts many of our youth to opt for a science-oriented activity?

One obvious answer to that question is that some youth are naturally attracted to involvement with science-related activities. Some are "loners" who have been performing experiments with chemistry sets or with science kits and who see a science club as a way of breaking out of their isolation. Others have been mesmerized by the *Mr. Wizard* telecasts; still others are in awe about the science and technology which make space travel possible. And some are the "joiners" who relate well to their contemporaries in a science/social context, namely, in the science club. A further reason to set up science clubs exists because of the intrinsic value in the dynamics of well-organized groups.

In addition, many youth need an audience with whom they can exchange ideas. These can be adults or peers. While not all adults need be authorities, fellow students enjoy the equality of their relationships as they share their interest in science. The club enables its members to receive feedback without specialized jargon. Science teachers enjoy exchanges at meetings and can understand the need to build socialization and recognition in the peer groups.

Also, a well-run club with elected officers, a written constitution and bylaws, and a series of annual activities can introduce students to democratic processes. If the direction of the group is self-determined, leadership emerges naturally with division of responsibility happening in time. The result can be that the group reaches important decisions through democratic means.

Multitudes of students have joined clubs that allow individuality and provide a forum where ideas can be tested. The potential for students' growth is great in an atmosphere that offers both science and socialization.

So there are several reasons why students join science clubs: Clubs enable individuals to be heard; they provide opportunity for leadership, for individual and group activities, and for recognition. Science clubs make peer support available, particularly for group projects, and make achievement more visible, more public, particularly when projects are displayed. Clubs offer an opportunity to meet others, and, perhaps, to make new friends.

Of particular interest to science-minded students are opportunities to meet scientists, to tour industrial facilities, and to travel to county, state, or national events. While many students simply enjoy the competition, clubs do not mandate it, tending more to encourage cooperation and collaboration.

But what of the students who appear to have a somewhat smaller degree of interest in science? We must at least try to capture their interest. A science club can create a forum that encourages discovery of interest where it did not seem present earlier.

Leonard M. Krause teaches science at William Penn High School and advises local and regional science clubs in eastern Pennsylvania. He contributed to NSTA's first Sponsor's Guidebook.

JOIN—TO COMMUNICATE FINDINGS, TO SHARE ONE'S WORK

John A. Blakeman

Many science textbooks begin with a description of the "scientific method." Few texts (and perhaps not all science teachers) recognize the essential importance of sharing one's research. Collecting background data, forming and testing an appropriate hypothesis, and so forth are all important, and students should know about them, but the entire "scientific method," as usually described, is virtually useless unless it also includes the process of communication.

Scientific data, discoveries, hypotheses, and theories are of little use to the scientific community unless they are shared with the members of that community. A scientist who conducts significant research but fails to share his results with his peers has, in an essential way, failed. The stereotype of a mystical figure, shrouded in a white lab coat in some dark, steamy laboratory, persists as the archetype of the practicing scientist, and any representation that the scientist merely collects and proves data only reinforces this notion.

In fact, modern scientists are usually dynamic, open, social characters. If they are to succeed, they must prominently share their thoughts and findings; they do so in a number of formats with which science students should have some experience. If scientific literacy includes a knowledge of, and experience with, the actual activities of modern scientists, then it is essential to teach science communication skills. Scientists share their research in two ways—orally at scientific meetings and in writing in scientific journals.

The primary vehicle for the transmission of scientific knowledge is the journal article. Students should learn how scientists write these articles and get them published and distributed. Most students don't understand how scientists learn of recent progress in particular fields. For all that many students know, scientists get their information from textbooks, just as do students. Students need to know about journal articles and their importance.

Likewise, students should become aware of the importance of scientific meetings, conferences, and symposia, where information is verbally shared and transmitted. Students should learn about how a scientist presents a paper, where a researcher stands before a group of peers and offers findings in an organized and persuasive manner. The interplay of thought in the subsequent question-and-answer period is equally important.

Perhaps the best way for students to understand these important concepts in modern science is to do them themselves. An involved, open-ended laboratory experience could be written up as a scientific paper, complete with an abstract, text, citations, reference list, and acknowledgements. The student could then present the paper orally to peers in the classroom, who would then probe with questions concerning the research.

Another method of communication used often at scientific meetings is the poster session. Here, scientists present their work written on panels of a large display board. Usually, the researchers stand beside their posters to receive and reply to questions from colleagues. Robert A. Day, author of *How to Write and Publish a Scientific Paper* (third edition, 1989), calls this an example of the "trickle up" effect from science fair projects (personal communication, 1989). (See page 53 for more information on poster projects.) Viewers can examine these, much as they might read museum displays. Having a poster session for a school science department open house is an effective way for students to communicate to parents and other visitors what has been happening in the classroom.

A visit to a college or large city science library can also be enlightening. Help students find the page (often the inside cover) in a current scientific journal where instructions for authors appear. Talk with a science librarian about appropriate scientific journals to examine. Most informative would be a description of a meeting of an Academy of Science. Most states hold annual meetings of their academies of science at which science teachers, with prior registration, are welcome.

Currently at work in the Perkins Local School District (Sandusky, Ohio), John A. Blakeman has been teaching biology at the middle and high school level for 20 years; he also does research on tallgrass prairies and birds of prey.

JOIN—TO BUILD CONFIDENCE, TO IMPROVE SKILLS

NSTA INTERVIEW WITH NICHOLAS D. FRANKOVITS, SCIENCE TEACHER, SPRINGFIELD HIGH SCHOOL, AKRON, OHIO

NSTA. Wearing the hats of both science instructor and academic coach at Springfield High School, you have chalked up an impressive record for entering winning students in science competitions. How many winners have there been from your school and over what time period?
FRANKOVITS. In this semi-rural system that sends approximately 30 percent of its students on to college, we have seen many students succeed in competitions at all levels. In the nine years that I have been academic coach, we have produced 4 international, 68 national, 55 state, and 30 regional winners in various competitions.

NSTA. The next question is obvious: What's your secret?
FRANKOVITS. Yes, that is the question we are always asked. My technique, which I also use in my teaching at the University of Akron, is simple and direct. And it seems to work. It is grounded in my belief that the best way to instill confidence, to motivate, and to develop appreciation for education is through academic competitions.

NSTA. Will you describe that technique?
FRANKOVITS. Each year I begin by restructuring my students' ideas and conceptions about how they perceive the sciences. Students envision the scientist, clad in a white lab coat, working in a laboratory filled with highly technical equipment. They equate the scientific world with hard facts, complication, and sophisticated equations. But in reality, many significant ideas and inventions originated in humble surroundings, as did the creation of "velcro." Who would have thought that this invention took place in the woods...not in a laboratory, but in the woods? As a man was walking his dog, he noticed that burrs attached themselves to his wool pants by tiny hooks. From this observation "velcro" was born.

NSTA. That sounds wonderfully simple.
FRANKOVITS. Next, I explain how many creative and imaginative people are not recognized by their contemporaries. For example, a

newspaper editor fired Walt Disney because he had "no good ideas", and Thomas Edison's teachers told him he was "too stupid" to learn anything. The list goes on. My classroom walls are designed to convey only positive thoughts through the numerous quotations that line them, such as "A success is one who decided to work and succeed. A failure is one who decided to succeed and wished." It is at this time that I introduce my classes to the "Wall of Fame," an area where I keep every award my students have ever won and every newspaper clipping that has been written concerning their achievements.

NSTA. That wall must make a strong impression on new students of science. How do they react?
FRANKOVITS. The Wall of Fame is a collage of discovery and accomplishment, and each year it continues to attract new students and generate that creative urge so necessary to the development of their potential. And, to the students' surprise, I explain that 80 percent of these award winners classified themselves as merely average students, not really capable of achieving this type of success. Not everyone will win, I go on to explain, but, in fact, in a sense all *do*, for the reputation of the school is enhanced and the name Springfield High School gains prestige. The high regard in which the school is held can even help our graduates be the successful applicant among many in line for the same job position.

NSTA. Do your students ever acknowledge that they understand the value of science competitions?
FRANKOVITS. I have asked present and past students what benefits they have gained through their involvement. I hear lots of responses— from building confidence, developing an appreciation for the value of education, understanding and admiring the scientists who have contributed and are contributing in the past and present. The students' ultimate feeling that they themselves can contribute and help to mold the future of this country is, I think, our greatest accomplishment...This is my motivation.

NSTA. What motivates the students? Is it the work in science that they do for the competitions, or is it the competitions themselves?
FRANKOVITS. Unfortunately, throughout many students' educational experience, they hear this statement from their teachers, "Someday you will use this valuable information that we are learning today." So year after year, students wait...for that magic day. Sometimes it never comes. Personally, I view much of our American educational system as being like football practice without games. How much enthusiasm and dedication would you get from high school

football players if you told them, "We will practice very hard each and every day after school…but…we will never have a football game"? Would you really expect dedication and enthusiasm? Of course not. Students need a game to play in which they can employ the skills being taught in each discipline in our schools today—mathematics, English, etc…Competitions are such games. And they work! I have given lectures and workshops across the United States explaining how to organize and launch a successful program. Not only has this approach worked for me, but now it is working for other instructors in the sciences, as well as other disciplines.

NSTA. What do you consider the most vital element of a successful competition program?
FRANKOVITS. Teachers are at the nerve center behind successful competition programs. Teachers must act more like coaches than like traditional classroom instructors. Coaches develop talent; instructors—inadequate ones—deliver fact and function. In addition to my regular classroom assignments, I serve as the school's academic coach. Although the position demands after-school sessions and weekend time, I am happy to be in it. Through out-of-school contact, I have often been able to develop that essential one-on-one relationship with the students. Here is where I am able to assess their latent talents.

NSTA. What sort of competitions would you advise a teacher who is just starting to organize a science competition program to enter?
FRANKOVITS. If you are considering launching such a program, begin with district and regional competitions as an introduction for not only you but also the classroom. And remember, we need to give students a reason to practice.

COACH AND TEACHER FRANKOVITS' STUDENTS AND PLAYERS SPEAK

ELEXIS A. BENCZO. I was first a student of Mr. Frankovits as a high school freshman, which was the year I became involved with the Space Science Student Involvement Program. I remember him devoting a portion of our class period to describing the competition. He made the program sound so exciting and rewarding that I couldn't help but be immediately interested. Mr. Frankovits has such a genuine enthusiasm for the Space Science Student Involvement Program that students such as myself can't help but be motivated to become involved in the competition.

I was fortunate enough to be selected as a semi-finalist that year, and from that moment on, my head was "in the clouds," and I vowed that one day I would be a national winner. During the following four years, Mr. Frankovits encouraged and supported me, offered suggestions and criticisms on my proposal, and raised my spirits if I felt hopeless. He never once let me believe that I could not reach my goal of becoming a national winner. After winning the semi-finals as a high school senior, I realized that not only did I want to win the nationals for myself, I wanted to win for this man who had given me so much of his time and concern. I wanted him to be proud of me.

Thanks to Mr. Frankovits, I have developed a strong interest in the United States space program. I remember so clearly the day of the Challenger disaster. It struck me as a deep, personal blow. I had been working on my Space Science Student Involvement Program proposal that very week and thus felt a very special interest in the shuttle. When the shuttle returned to space in the fall of 1988, I sat with my fingers crossed and my heart pounding as I watched the lift-off on television in the lobby of my dormitory at Mount Union College. In fact, I am currently considering a physics/astronomy major at college. I would love to be involved in the space program someday, and much credit would have to be given to Mr. Frankovits if my life takes that direction.

It's difficult to express in words how Nick Frankovits has touched my life. For four years he has been my teacher, advisor, and friend. He has served as a role model for me, and I greatly admire his positive attitude, his genuine enthusiasm, and the care and concern he shows for his students. Mr. Frankovits has instilled in me things that could never be learned [only] in books. I gained confidence after I realized that I *could* successfully complete the seemingly impossible task of writing a scientific paper. He taught me to believe in myself, and to believe that my ideas were worthwhile and of significance. But perhaps most importantly, Nick Frankovits taught me to dream. He showed me that the whole world (in fact, the whole universe!) is open to me, and if I follow my dreams and work hard, I can accomplish anything.

JAMES SKALSKY. I am currently a senior, age 17. I entered a national competition [by making] a safety device for ladders. This device alerts people to the potential hazard of unlevel ladders. Placing in a national competition, or any competition, is a great feeling of accomplishment. My device helped me to feel acknowledged and has headed me towards the science field. I like the challenge of competitions and the feeling of success. I dislike an application [with] rules that [can be] overlooked. This can lead to inconsistent guidelines...

Competitions are a valuable experience for all who enter them.

BOB A. SUTTER. I am freshman at Springfield High School. Although I have not won a national competition, I have just submitted a paper to NASA on monitoring the depletion of the stratospheric polar ozone layer from space with an image spectrograph. Not only has this idea encouraged me, but I feel that its potential, with the possibility of winning, will inspire me to win many other competitions.

Not only has this paper contributed to my success, but if the ozone layer is photographed, and a successful diagnosis of the culprit of its depletion is found, a remedy may be found and people throughout the world will be safe from the harmful ultraviolet light of the sun, which leaks through the ozone hole.

DAWN BALDT. As a student of Mr. Frankovits, I feel his effectiveness as a teacher comes from his drive to obtain knowledge and his willingness to help others succeed.

Mr. Frankovits spends a large portion of his time helping (sometimes very unmotivated) students enter their (usually very complicated) science projects into different contests, such as the Space Science Student Involvement Program, Duracell NSTA, East Ohio Gas, and the State Science Fair. For he knows that, with these accomplishments on students' records, they will have an advantage over others to get the job they want. Most of the winners come from one school in the Springfield area, with one teacher behind it all, Mr. Frankovits. This is a large job for one teacher, considering there were 55 entries in the 1983–1984 school year. These science contests, along with his commitments at Akron University, and his job as a science teacher, cut into...his sleep.

Mr. Frankovits does a superior job, not just teaching kids but making them want to learn and want to get more out of what he, other teachers, adults, and life have to offer. He prepares students with knowledge they will need to have in the future, such as [that which will be tested] on important exams. He tells us what certain things *really are* in comparison to our fictitious beliefs. He teaches us a wide variety [about] many different areas, making our knowledge span broader and making us competitive with other schools and other countries. He makes us aware of life's surroundings.

Mr. Frankovits's objective as a teacher is to get the most out of one of the world's most powerful resources not nearly used to its full capacity, the human brain.

JOIN—TO LEARN

Gene Kutscher

There are numerous competitions available to students who have
completed an independent research project in science. The benefits of
competing are many; they include building self-confidence,
developing critical thinking, and enhancing organizational skills.
Paradoxically, however, students are often reluctant to prepare their
work for presentation because they lack confidence not only in their
results but also in their ability to display their research. The positive
encouragement of an interested teacher-mentor is of paramount
importance to strengthen skills enabling students to compete
successfully.

Students should never be required to compete, yet there are
substantial reasons for teachers to persuade most student researchers
to do so. Generally, competitions require students to submit the results
of their science research in writing and/or through oral presentation.
Both methods build the kinds of skills and confidence that remain
with the students for life, including the ability to conduct critical
analysis, to persist, to develop written and oral communication
techniques, and to recognize that questioning and defending results
promotes intellectual growth as well as scientific progress.
Competitions also permit the broad sharing of work and ideas that are
necessary for the growth of science and vital both to young
researchers and to scientists.

As a first step, it is important to build students' confidence in
their results—whatever those results may be—because this response
cuts to the heart of true scholarship. For example, unanticipated or
null results, properly obtained, must be viewed as a cause for
exploration rather than sorrow, shame, or confusion. Some of the
finest research can verify a null hypothesis or can proceed in
unanticipated directions. Furthermore, eliminating a false trail is an
important part of the scientific process. The brilliant work of
Thorndike, Kennedy, Michelson and Morley clearly demonstrates this
point. They invented wonderful instruments and techniques while
trying in vain to find differences in the velocity of light perceived by
different observers. Ultimately, Einstein presented his (then radical)
theory of relativity to explain their surprising results.

Next, the development of requisite skills is a matter of having the proper knowledge, good advice, and actual practice. In order to present their research, students should write a paper following a basic script, no matter what the subject. The paper (whether written or delivered orally) should have an appropriate title, followed by an abstract of the research, a background or introductory section, a statement of the project goals, a working hypothesis, a description of procedures and materials, a compilation of data and important observations, a presentation and analysis of the results, suggestions for additional research, and a bibliography. (See John A. Blakeman's paper, page 53, for more details.) All must be organized in a style appropriate to a scientific paper and contain materials presented in accordance with scientific ethics.

Students usually need assistance with each of these steps. Three areas require special attention. The first is formulation of a good problem statement. Once this is narrowly defined, the hypothesis and methodology are more apparent.

The second is analysis. Students need assistance with graphic presentation, use and interpretation of statistics, and visualizing the relationship between the outcome of their research to the overall field of science. Teachers should discuss each of these aspects of the report with their students, allow the students to develop the sections in writing (when possible it is helpful to use word processors), review the results, and suggest improvements.

The third is the need to follow rules. Most competitions will not judge material unless it follows all of the criteria for entry, including matters of style and student identification. Therefore, both teacher and student should scrutinize the final paper for accuracy of format.

When competitions include oral presentations, students must rehearse, *not* merely practice alone. A subsequent presentation before a group of peers is most helpful for building confidence and suggesting improvements. Few students are comfortable with oral communication, and the supportive advice of a friendly teacher is treasured. Using index cards, students should record and memorize key points of the paper. They should also practice making reference to accompanying physical projects and/or audio-visuals (such materials should be high quality). Next, students should rehearse the entire presentation several times, keeping in mind time limitations, the location of the talk, the type of audience (single judge, small group, auditorium), and the judging criteria. Small points are important: appropriate clothes; a pointer for highlighting areas of projected materials; speaking from, rather than reading, the paper.

The close support of a teacher-mentor can result in a productive experience, both personal and academic, for student researchers who enter competitions. Thus, entering competitions can build invaluable life skills and confidence, can foster a special kind of learning, and can reinforce positive feelings about science.

Gene Kutscher, who has taught science for 22 years, chairs science and coordinates science research at Rosyln High School (New York). He is the author of Physics Research Activities *(1989) and* Creative Science Activities, Grades 5–9 *(1990) (both from Alpha Publishing in Annapolis, Maryland).*

PART II:

Getting Involved

Participating in
Science Activities

SCIENCE INTEREST PROGRAMS

John A. Blakeman

There are many kinds of valuable extracurricular science opportunities. One that has worked well at Perkins High School in Sandusky, Ohio, is the Science Interest Program, a series of science activities throughout the school year that stimulate interest and provide activities not normally possible in the classroom. Traditionally, science clubs have served this purpose, but sometimes they are deflected from science experiences into electing officers, collecting dues or raising funds, conducting meetings, forming committees, and wrestling with parliamentary procedure! While these activities can become central—and are often valuable—at Perkins, science faculty arranges instead science activities with maximum student participation.

Perkins' Science Interest Program is primarily a department-sponsored series of after-school field trips to places in the community where science happens. When we asked a number of local industries if they did research and development or had quality control labs that we could visit, virtually all were delighted to show a group of interested students how science happens at their facilities. Teachers speak with the particular lab's director before the visit to suggest subjects that might be covered. Students have been astonished to discover the amount of science done in factories right around the corner from the school.

Trips to science and natural history museums are also effective, but again the experience improves with preparatory communication between faculty and museum staff, an interaction that often gets us behind the scenes to talk with the researchers.

Or, we visit local natural areas. A hike through a woods or natural ravine with a knowledgeable faculty leader who points out life and Earth science features is always popular.

Other useful and informative activities are visits to nearby college science departments. Unfortunately, many high school students, especially freshmen and sophomores, have never visited a campus science laboratory, nor have they talked with college science students or professors.

When students occasionally express interest in conducting their own research projects in our labs after school, this independent work comes under the purview of the Science Interest Program.

Our Science Interest Program has no members, only participants. To participate, students need only preregister with the sponsoring faculty member. Many students participate in only one or two events each year, choosing activities in which they are particularly interested. All activities are open to the entire student body. Events are posted on a calendar in all the science rooms and are announced, often humorously, on the daily morning announcements. These announcements enable everybody in the school, from the principal to the youngest student, to know that science is alive and well, both within and beyond the classroom.

For John A. Blakeman's biography, see page 6.

IN-SCHOOL SCIENCE CLUBS

Peter G. Bruecken

The first step in starting a science club is persuading the school's administration to accept the club's goals, including, perhaps, some of the following aims:

- to stimulate interest in scientific activities
- to broaden the academic endeavors of the school
- to help students realize their scientific skills in relationship to other students in the scientific community
- to illustrate the role science plays in everyday life
- to include the school in community scientific activities
- to include the community in school scientific activities
- to promote scientific thinking as a way of looking at the world and not merely as a way to manipulate it
- to make scientific activities *fun* instead of just another set of courses

Affiliations with national and state organizations can increase interest in the science club and offer the opportunity for students to expand their exposure to a wider range of challenges. National organizations offer competitions, tests, meetings, camps, publications, and specific information that can be of interest to science students in your school. These organizations also expose your club to people outside your community who share similar interest in scientific activities. Your club can subscribe to publications of interest not readily available in your particular area. These broader horizons can take your club beyond its physical boundaries as well as bring outside influences into your school.

Affiliations with science-teaching organizations can be a wellspring of opportunities for the science club. These organizations can share meetings, advertisements, competitions, articles, and reliable projects with a large group of people with similar interests. Through their publications and meetings, these organizations can expose students to enterprising activities and thereby improve the club's image.

THE SCIENCE DEPARTMENT'S CONSIDERATIONS

A science club can positively affect the science department. It can encourage members to use the club to generate interest in their courses by involving club members in their particular activities. Time or restrictions can mean that many opportunities for students, arising in specific disciplines, go to waste. Extracurricular clubs can extend and take advantage of such opportunities. The department can use the science club as an outlet, providing a unifying influence on extracurricular activities and making them available to more students. The members of the department then have an easier job motivating their students. An increase in science course enrollment may even occur, involving more students in science.

ADMINISTRATIVE CONSIDERATIONS

After setting the initial goals of the club, have them evaluated by the administration and then work with other faculty to help meet them. At this point, many administrators will be fundamentally uninvolved in the club, but their early participation may help establish initial guidelines and avoid possible future misunderstandings. Even if the administration's involvement is minimal, keep administrators informed about the club. Then, if it achieves something outstanding, the administration will not be unaware of the club's existence when it is complimented. Club officers should always give a report of planned club activities to the administration.

After the goals of the club are formulated, club members should prepare a budget and a meeting and activity schedule, including transportation details. Confront these logistical considerations at an early stage, before problems arise. It is important to plan some reoccurring annual activities, like competitions, so that students will have a particular science club event to look forward to every year. Flexibility is important to meet the changing needs of new students each year, but some consistency is needed to keep the club going through the years when enthusiasm wanes.

Early planning, as outlined here, will enhance the club's relationship with the administration and facilitate a good working climate. It will also enhance the reputation of the science club moderator.

After goal-setting, the next step is to write the club constitution. The constitution should provide for election of officers, membership eligibility, meeting protocol, and purpose of the club. The constitution should be in harmony with school policy and provide for changes in policy. For example, students on academic probation should not be allowed to participate in any science club activities. The constitution should spell out procedures to follow in such situations, especially in cases where the student involved is a club officer. The constitution should be flexible enough to permit change, but specific enough to make the club effective and consistent. (See the sample constitution in the next section.)

It is important to consider club finances at the outset. The club will need money for affiliations, transportation, and competition entries. Funds may become available as a lump sum or may be raised from a variety of sources. The science club may raise funds from dues, contributors, or local businesses and industries. The school may have funds available for science club expenses. If so, the amount should be established in the spring of each academic year. Most schools have a budget deadline, and the science club budget should be submitted before the deadline arrives. The budget should involve club members and, of course, be acceptable to the administration. A good budgetary approach is to attempt to match funds from many sources.

Students tend to be more involved if they are raising some or all of the money themselves. Student projects can add to the interest and involvement of the membership. All expenditures should be reported to the school administration to keep it aware of club activities and accomplishments.

STUDENTS' CONSIDERATIONS

During the club's formation stage, collect a core of students who are actively interested in it and try to get their involvement in developing its constitution. Try to establish at least two reoccurring annual activities. Then, students will always have those activities on which to plan, and the school community will look forward to them. Such annual activities will not be the only ones that matter, but they will serve as consistent projects to establish continuity from year to year.

Have an election of officers, but make sure the candidates know what is expected of them. At the nomination meeting, make sure that the description of each officer's position is read and the individuals who are running know their responsibilities. The job of club moderator, for example, is to encourage, provide opportunities, and handle logistics. It should be clear to the members that this is their club, which they should make the very best. Frequently, students are not aware of the influence they have on school activity clubs and what their potential is for making it a success.

Elected club officers help to free teachers from some of the work involved in operating a club and give other students a feeling of accomplishment when things go right. If the teachers are liberal with compliments, students will be motivated to do a good job. They should have fun working in the club and be rewarded with a sense of accomplishment.

Students should get proper recognition from all the appropriate places. They should cooperate with the school paper and the local media and assist in preparing and disseminating school announcements. Cooperation can take many forms. Sometimes it is up to the club members to take club pictures, write club articles, and contact the appropriate people to obtain proper recognition through publicity. Such promotion is an important part of keeping the science club going and encouraging future members. Most of this promotional work should be done by the students, because they will be the main beneficiaries of recognition.

Again, the moderator's main responsibility is arranging logistics, keeping order, offering encouragement, and giving guidance. If the moderator leads student members yet allows them to make their own decisions, the students accomplish more tasks independently.

The parents of the students can and frequently do get involved in the activities of the science club. Parents can have a positive influence on the activities of the club and help motivate students to do well. The parents who get involved are usually the ones who have objectives similar to those of the members. If the moderator can channel parental involvement, the club will be better as a result.

COMPETITIONS

Probably the most popular activities of science clubs are competitions. These range from local science fairs to the international olympiads. Many colleges and universities hold a "science day" offering some kind of competitive event for high school students. There, students of different schools mix in activities that help identify skills and motivate excellence in science at the high school level as determined by student peer groups.

Whether the contest is a test, a research event, or a manipulative hands-on project, your students will be competing with their peers, a fact that helps them identify with the other contestants. For some students, the awareness of other groups will help them feel better about themselves. By bringing different student groups together, competitions can engender mutual respect.

Along with competing among themselves, members of the science club can have a profound effect on younger students. Running a competition for students in the lower grades is a good way to secure the future of a high school science club. Science club students have a significant influence on students in lower grades: in running a school-wide competition, the science club students serve as role models for the young. Club members, by showing interest in younger students, have a positive influence.

The experience of running a competition themselves will enable science club members to perceive contests in a different light, sharpening their perceptions and enhancing their own performances in science club contests.

TRIPS

Field trips to local places of scientific interest are helpful to students' concepts about the use of science in their immediate worlds. Many ordinary things like water, air, and electricity, can be good subjects to investigate. Tours often motivate students to look further and help them learn science by illustrating the *need* for scientific applications in things that some students take for granted. Often, students are amazed at the degree of sophistication involved in apparently commonplace concerns. Tours can help display the vast dimensions of science affecting our immediate environment.

Tours can also be the beginning of a bond between the science club and a local industry. Besides the interest and learning gained by seeing the science behind everyday realities, members may find business support for the science club or for a project. Many utilities, for example, are willing to work with local school clubs. But the industry needs contact with the school to stimulate interest. If a business and a club make this contact, it could result in an significant contribution to the community.

GUEST SPEAKERS

Guest speakers are another vehicle that can bring experiences from the outside world into the school. Such speakers broaden the interest of students by introducing them to some aspects of science that are not otherwise available. A guest speaker can help to humanize science through direct contact, adding to students' exposure to science from books, labs, equipment, and courses. In addition, the speaker may choose to reach out to the school in other ways. Guest speakers have an effect on the school, and the school can have an effect on them.

IN SUM

An effective science club can enhance the learning of science in many ways. By involving parents, school officials, teachers, the community, and students in extracurricular science activities, high school science can be a more meaningful subject of study, even a way of thinking. Science activities and organizations enable all participants to become more aware of their own scientific gifts and develop them in an organized fashion. In schools, club-based competitions can increase the level of achievement in science; tours and guest speakers can broaden the outlook on science; and high school involvement with students in lower grades can make science more than "just another class" to those students who choose to expand their scientific endeavors.

Peter G. Bruecken has been teaching high school physics and physical science for 15 years, currently at Heelan High School, in Sioux City, Iowa.

SCIENCE CLUB CONSTITUTION*

ARTICLE I. Name
The name of this club shall be..

ARTICLE II. Purposes
Section 1. To learn about science, science careers, and the opportunities, responsibilities, and important role science plays in our society and the world.

Section 2. To explore the interests and abilities of personality, character, and leadership that are essential to scientists.

Section 3. To cultivate in members the qualities of personality, character, and leadership that are essential to scientists.

Section 4. To learn how and where scientists receive their training, its cost, the scholarships available, the number of years required, and the academic prerequisites and standards.

Section 5. To study the lives and the influence of great scientists.

Section 6. To stimulate interest in and interpret science to others.

ARTICLE III. Officers and Duties
Section 1. The president shall officiate at meetings and see that the work of the club goes forward.

Section 2. The vice-president shall assist the president and act in his or her absence. He or she shall serve as chair of the program committee.

Section 3. The secretary shall keep an accurate list of members, a record of their attendance at meetings, and a written log of the club, including the minutes of business meetings. He or she shall carry on all club correspondence.

Section 4. The treasurer shall collect dues and keep the club accounts.

Section 5. The director of public relations shall send news releases to the media to inform the public of club activities.

Section 6. The historian shall keep the history of the club and the scrapbook.

Section 7. The librarian shall care for club materials and make them available to members.

*Reproduced with permission from the *Sponsor's Guidebook*. Copyright © 1971 by NSTA.

ARTICLE IV. Qualifications and Duties of Sponsors

Section 1. The sponsor shall be a teacher approved by the principal.

Section 2. The sponsor shall guide the club in all of its activities.

ARTICLE V. Membership

Section 1. Membership is open to any student in grades 6–12 who is interested in exploring science as a career and who has the high qualities of character, personality, scholarship, and leadership essential to a good scientist or science teachers.

Section 2. Two consecutive unexcused absences from regularly scheduled meetings may cause loss of membership.

Section 3. Members must maintain a minimum GPA of 3.0. The scholastic and citizenship standing of each member shall be reviewed periodically.

ARTICLE VI. Membership Dues

Members shall pay annual dues of ....

ARTICLE VII. Meetings

The club shall meet regularly at ....

ARTICLE VIII. Elections

Officers shall be elected annually by secret ballot.

ARTICLE IX. Quorum

A simple majority of the members shall constitute a quorum.

ARTICLE X. Committees

Section 1. The president and sponsor shall be ex-officio members of every committee.

Section 2. Committees shall include program, social, public relations, membership, finance, and service.

Section 3. The vice-president shall chair the program committee.

Section 4. Chairs of the other committees shall be appointed by the president.

ARTICLE XI. Amendments

This constitution may be amended by a two-thirds vote of the membership at any regular meeting. Written notice of the proposed amendment must be filed with the secretary and presented at the meeting prior to the one at which a vote will be taken.

SCIENCE CLUB SURVEY RESULTS

In October, 1989, *NSTA Reports!* ran a survey to be completed by any interested member or advisor of a science club at any level. See part IV (appendixes) for examples of completed forms. Seventy-nine surveys were returned, a number of them accompanied by supplementary materials. A summary of the results follows.

THE NUMBERS

In response to questions calling for numerical responses, NSTA received the following ranges of answers:

Table 1			
Seventy-nine Science Clubs at a Glance			
	Elementary Junior High	Middle School	High School
Surveys returned	10	17	52
Pupils in school	10–700	18–1,200	15–1,500
Years in existence	1–5	1–20	1–50
Number of members	13–25	7–198	10–300
Number of meetings	0–30	0–75	0–150
Hours advisor spent	0–50	0–150	0–1,000

Although 79 returns constitute an extremely small sample, the responses are of interest. Some further subtotals, again admitting the small sample, also bear quoting, however.

- In 10 elementary schools, 10 advisors spent 202 hours serving 177 club members (out of 2,823 pupils). This averages to over 20 hours per year per teacher.
- In 17 middle and junior high schools, 17 advisors spent 1,001 hours serving 750 club members (out of 7,110 pupils). This averages to almost 59 hours per year per teacher.
- In 52 high schools, 52 advisors spent 4,174 hours serving 2,351 club members (out of 46,041 pupils). This averages to over 80 hours per year per teacher.

So running the science club at any level requires a substantial outlay of effort.

THE ACTIVITIES

Most clubs at all levels went on field trips, did experiments, and worked on science fairs. Some elementary school students also discussed books and held a "science magic show." A good number of club members from middle schools, junior highs, and high schools also heard guest speakers, raised money, and worked on recycling or other nature and conservation projects. The early adolescents in one club also grew and propagated plants and in another attended a flower show.

The activities of the 52 high school clubs also included a good deal of work on competitions beyond school fairs; three respondents specifically mentioned work in the Science Olympiad. One club worked in environmental studies (specifically, like another, in geology); while eight clubs were active in conservation and environmental projects. One club did school service; members of another taught elementary students, held a plant sale, and went on a retreat weekend. Besides these activities of a rather general nature, individual clubs participated in more unusual projects.

One high school club, for example, developed a "women in science" project. Another held a "mole day competition" and awarded a $500 scholarship to a graduating senior. One California science club with 80 members is the "largest in our school...and still growing." Many clubs take advantage of the area in which they are located; thus, a suburban New York club visited the Bronx Zoo, a Florida one learned about manatees, and another California group, the "Swamp Stompers," takes care of and builds ponds and creeks.

OUT-OF-SCHOOL SCIENCE CLUBS

Mark A. Wagner

What do the Space and Rocket Center in Huntsville, Alabama, the Fontenelle Forest Nature Center in Bellevue, Nebraska, and the Museum of Science and Industry in Chicago all have in common?

In Huntsville, home of the famous "Space Camp" where children can experience a simulated astronaut training course, enrollment demand far exceeds capacity. So, to ease the disappointment, the Space and Rocket Center's "Space Club" provides space-related activities and materials for students who cannot get into the camp program.

In Bellevue, aspiring young naturalists and nature photographers in grades 5 to 12 meet once a month to enjoy the Fontenelle Forest Nature Center's three naturalist clubs.

Numerous other science clubs are thriving across the country, including the Discovery Place in Charlotte, North Carolina, the Franklin Institute in Philadelphia, and the Science Center in Des Moines, Iowa. All three have student clubs that extend science investigation into a year-long, extracurricular activity.

In Chicago, the Museum of Science and Industry's Science Club offers hands-on science club experiences to the approximately 40 members in grades 7 to 12 accepted on a first-come, first-served basis. Members, who pay a nominal fee, meet every other Saturday morning during the school year in the Museum's Seabury Laboratories.

FOCUS ON CHICAGO

I have for five years served as coordinator for this club. As such, I plan and run the meetings, with an assistant who also helps with preparation and implementation of science activities. In addition, the club has several adult volunteers, about three of whom attend any given meeting. Some of these volunteers are elementary school teachers who see the club as offering a way to learn new skills and activities that they may incorporate into their own science teaching.

A typical meeting begins with all of the club members sitting around tables in one of the labs. I am in charge of the business portion of the meeting, welcoming the members and asking them to introduce any guests (many current club members first attended as guests). Then, I make announcements, ask the members for suggestions for future project topics, and get general feedback about club activities. The business meeting lasts for about 15 minutes.

Next, the members break into two project groups. Those in each group work on the different topic areas they selected at a prior meeting. Typical areas include field biology, electric gadgetry, electric circuit design, balsa-structure building, air-track construction, custom circuit-board etching, and kite construction and flying.

The club year is organized into three activity cycles of six meetings each. The first five meetings of the cycle are devoted to projects centered around a unifying topic. The sixth is a sharing session during which each group plans and makes a presentation of its activities and finished products. During the business portion of this meeting, the club members sign up for their new group for the next cycle of activities.

During the weeks preceding the sharing meeting, the club staff and volunteers brainstorm to find topics for the next cycles. Club members also provide ideas at business meetings. Typically, members suggest broad areas of interest such as chemistry, radio-controlled vehicles, or electronics, but leave the specifics to adults.

The aims of the club are to

- foster long-term interest in science
- improve members' skills and knowledge in scientific and technical areas
- increase the likelihood that members will pursue careers in scientific or technical areas, including science teaching

These broad aims shape the more specific objectives of the project groups, which are formed during the three cycles of the club year. However, members do not decide on specific group topics and objectives until two or three meetings before a new cycle begins. In general, there is little repetition of specific activities, although members often address general topics several times through new approaches.

Electronics is a popular topic and has been taken up often through different activities. In the last three years, groups have worked with robotics, computer interfacing, electric gadgetry, circuit design, and digital electronics. Many of the skills fostered by the activities in these groups overlap with different emphases. One of the challenges faced in developing group topics is how to provide activities appropriate both to beginners and to experienced members.

Most of the group activities center around construction or experimentation with little formal instruction. Formal presentation of scientific facts are kept to a minimum, and instruction often takes the form of coaching members on an individual or small group basis as they need help. Members often finish a cycle having raised as many questions about the topic as they have resolved. As it turns out, this facet builds continuing interest among many members, who await the reincarnation of a topic under a new banner to continue their exploration.

ONE PROJECT: CIRCUIT BOARD ETCHINGS

One of the unique features of the Chicago Museum of Science and Industry Club is its responsiveness to the members' interests. These interests, along with those of the staff and volunteers, have helped guide the club into exciting areas of science and technology not experienced by most students in typical classrooms.

For example, during the business meetings, several members expressed interest in constructing electronic devices. In previous cycles, certain projects had centered around electronics, and the club had accumulated a library of electronics materials containing hobby magazines, project books, technical manuals, and catalogs from supply houses.

In looking through one of the catalogs, I stumbled onto an advertisement for a circuit-board etching kit. I had heard of the etching process, but neither I nor my assistant had any experience with this technique. Both of us found the topic interesting. At the next meeting, I asked one of the more experienced club members if he had etched circuits before. He had not, but was also excited. So, I ordered the kit, and with the help of my assistant and one of the volunteers, we tried out the process.

After a little practice, etching proved to be within the capability of relatively inexperienced experimenters. I planned several activities using rather complicated circuits, but one of the volunteers pointed out that—like herself—many of the members were new to electronics and might profit more from simpler projects. She also wanted to know more about how the components function in a circuit, rather than just using them to assemble devices. I agreed that, while the focus of the new electronics group should be to learn the basics of circuit design and construction, the process of etching would add interest.

The group would learn about various components by building progressively more complicated circuits, each emphasizing one component and showing its properties. The first circuit would test the continuity of electrical connections. This circuit required members to select an appropriately valued resistor to protect a light from receiving too much current from a nine-volt battery. A resistor valued too high would reduce current flow below the threshold of the bulb, while one too low in value would let the light burn out. The next circuit used two light-emitting diodes to demonstrate how these polarized components allow current to flow in only one direction. The third circuit showed how a transistor cell amplifies a signal. The fourth project used capacitors to store electricity, which alternately flash two light-emitting diodes.

Members were given schematic diagrams for these circuits but designed the component placements themselves. This process led to much experimentation. To facilitate the connection of components, members first constructed their circuits on a temporary circuit board consisting of many sockets arranged into a plastic triangle called a breadboard. By using temporary connections, the members tested various arrangements to see if their interpretation of the diagram was correct. The breadboards also allowed easy modification and customizing of circuits. When the circuits were functional, members etched a permanent circuit board. The resulting circuit had copper lines connecting components soldered in place.

Members of the circuit-design group worked through these projects at their own pace. Some proposed and carried out variations on these projects. A few advanced members completed all of the circuits and selected other interesting circuits to experiment with from the club's "electronics library."

Completed projects become members' property. Showing their projects to friends, parents, and teachers at school reinforces members' satisfaction at completing rather difficult projects in technical areas.

LOOKING AHEAD

In the future, the club intends to develop chapters in Chicago-area schools and youth organizations. If the plan goes as expected, teachers and other adult group leaders would attend three-week summer training institutes at the Museum taught by the club's staff and student members. Participants would learn how to set up a club using activities developed at the Museum and would receive a kit of materials to implement a year's worth of activities in science and technology. The Museum chapter of the club would serve as a development center for activities to be used in other chapters, multiplying the effect of the staff and volunteer efforts in perfecting interesting, worthwhile projects for many students citywide.

I hope that, as the number of clubs in the Chicago area proliferates, the number nationwide will grow as well.

Mark A. Wagner has coordinated the Science Club at the Museum of Space and Industry for five years. In addition, he teaches science in the public schools in Chicago, currently in a magnet program for seventh through twelfth graders at Kenwood Academy.

Research and
Competition

PLANNING MAKES PERFECT: STEP BY STEP TO A GREAT SCIENCE FAIR

Nancy C. Aiello and Brian E. Hansen

If you are excited about your students having a hands-on, in-depth experience with science but are uncertain about how to begin planning a science fair, you can *relax*. Help is available, and the planning is easier than you think. In fact, Larry Farmer, math and science supervisor for Virginia's Loudoun County Public School system and organizer of eight regional science fairs, says, "The details of planning a science fair will fall into place if a teacher has enthusiasm for science" (personal communication, Aiello and Hansen, 1989c).

The excitement of a science fair project, Farmer says, is that students see science happen. They don't just get it out of a book. B. Blaine Soto, current dean and a former science teacher at Loudoun County's Broad Run High School, also likes the fairs because they give her an opportunity to get acquainted with her students personally, as she cheers and coaxes them in their search for answers to scientific questions (personal communication, Aiello and Hansen, 1989e).

Let's look first at what high school science fairs should and should not be.

The fair *should* be an opportunity to present research and conclusions about real scientific problems. It *should not* be a collection of exploding papier mâché volcanoes, plants murdered in dark closets, or colorful crystals grown in water. The judging of the projects should be instructional, an opportunity for young scientists to discuss their research with and learn from experts in the field. The fair *should not* be an art contest with projects having elaborate displays and little scientific substance. It *should not* be an experience requiring clenched teeth and a bottle of aspirin. It *should* be a worthwhile experience for students, teachers, judges, and the public.

GETTING HELP

Agencies, books, and videotapes are abundantly available to help plan and conduct a science fair. Several sources of help are listed at the end of this paper. Science Service, Inc. (1719 N Street, N.W., Washington, DC 20036), an organization that sponsors the annual International Science and Engineering Fair (ISEF) and administers the Westinghouse Science Search, is a gold mine of information. Science Service writes and publishes a number of useful documents specifically concerned with the ISEF but generally applicable to many other science fairs (1989a, 1989b, 1989c, 1989d, 1989e). One of its publications (1989c) is *Guide to the [most recent] International Science and Engineering Fair*, an annual publication containing information about the current year's international fair and several pages of suggestions about conducting a local fair. A second useful ISEF document is *Rules of the [most recent] International Science and Engineering Fair* (1989e). In addition to the rules, this publication contains forms that affiliated fairs should use. Single copies of these two pamphlets are available at no charge, and additional copies can be obtained for a nominal price from Science Service, Inc.

NSTA is another source of assistance. It has two pamphlets specifically dealing with science fairs—*Science Fairs and Projects Grades K–8* (1988) and *Science Fairs and Projects Grades 7–12*, (1988) made up of articles originally published in the 1980s in *Science and Children*, *Science Scope*, and *The Science Teacher*.

School and community librarians can also help you. There are many books on science fairs available, such as Maxine Haren Iritz's *Science Fair: Developing a Successful and Fun Project*, (1987). This book, written for the high school student working on a science project, may give teachers some useful suggestions about the fair. *Switch on to Science* (1986) is a motivational videocassette which presents ideas for topics in various categories. The teachers' manual lists over 400 additional science project ideas. You should also alert your local librarians that students will be searching for specific information on their project topics and ask them to order appropriate scientific journals and books.

Don't overlook the advice of teachers in nearby school districts. They may be able to give you advice about everything from the supervision of student projects to the names of possible judges.

TO JOIN OR NOT TO JOIN

After you collect information on planning science fairs, but before you put your sweatband on and get to work, you should meet with colleagues, your principal, and your district science supervisor to decide whether you want your fair to be affiliated with the ISEF. If your district is already sponsoring an ISEF-affiliated fair, you will probably want to follow the ISEF rules, so your school fair can feed into the ISEF-affiliated regional and state fairs. If there is no ISEF-affiliated fair in your district, then your decision is a little harder.

One advantage of ISEF affiliation is that the winners of your fair can go on to the international competition. Another advantage is that you may receive awards from numerous professional organizations and businesses that cooperate with Science Service. Some of these organizations will even supply judges for your fair.

Your school may not want to affiliate with ISEF because of cost. Affiliates must pay Science Service between $200 and $600 and must agree to fund the round-trip travel and living expenses for up to two fair winners and one adult escort, as well as shipping the project, to the international competition, which is usually held within the United States. Another consideration may be that you are unable to or do not wish to follow Science Service's strict rules. For example, the ISEF does not allow group science projects, and each sanctioned fair must have a Scientific Review Committee.

GETTING THE BALL ROLLING

The next step is to form a *science fair committee*. Draft everyone who is enthusiastic: science teachers, the science department chair, someone—maybe an English teacher—to take meeting notes and write news releases, someone—perhaps an art teacher—to advise students on displays, and parents to help in general ways and to serve as liaisons with the community.

One of the committee's first decisions should be whether participation in the fair will be mandatory or voluntary. Required participation results in a lot of students and teachers getting to know each other better. Moreover, students previously "turned off" to science may become interested in it as a result of the fair. Mandatory participation also provides the opportunity to make projects into class requirements and enables the teachers to better monitor them. It also results in an impressive, well-attended fair. Voluntary participation, on the other hand, besides being a policy recommended by NSTA, reduces the number of disgruntled students, parents, and teachers who don't see why humanities students have to "waste their time on science." Broad Run High School, for example, required participation for its first three science fairs but now has switched to a voluntary fair.

In the mandatory–voluntary debate, several intermediate positions, suggested by Joy Bauserman, a physical science teacher at Broad Run (personal communication, Aiello and Hansen, 1989a), include the following:

- Offer a science fair elective course, to allow interested students to work with an enthusiastic teacher.
- Require science projects for certain students. At Fairfax High School in Virginia, honors and advanced placement students must do projects. At a small high school, all juniors and seniors could be asked to participate, while freshmen and sophomores could be introduced to science projects through group projects and science demonstrations.
- Start a science fair club which meets during an activity period and is advised by a dynamic science teacher. The club would serve as a meeting place for students to exchange ideas and enthusiasm and to get help from the advisor.

A second important committee activity is establishing a working calendar. Set the date of the fair and work back to the present. Include dates for accepting project application forms, contacting judges, arranging for commercial and government exhibitors, and acquiring all awards. Remember when you are setting your fair date that if you are affiliated with or feed into an ISEF fair, you must meet its deadlines as well as your own. The names of students participating in the international competition must be sent to Science Service by mid-April each year.

Another advantage of an early date, possibly in February, is that you can make it occur in the lull between winter and spring sports. Because the gymnasium is usually free then, battles between the coaches and the science teachers over the use of gym facilities become unnecessary. Early in the planning process, the science fair committee should reserve the facilities for the entire day of the fair. A full day is necessary for project set-up, judging, public viewing, and the awards ceremony. A standard gym can accommodate about 200 projects. Another facilities option, especially if your gym is not available or your fair is on a Saturday, is to use classrooms. At the Virginia State Science Fair, projects are grouped by category in adjacent classrooms at the Loudoun Campus of Northern Virginia Community College. A third facilities option, one used successfully by Fairfax High School, is to hold the fair in your school's hallways. There are advantages and disadvantages to each arrangement. A gym is easier to set up, but can be pretty noisy when it contains 200 students and 70 judges. Individual classrooms are easier to manage during the judging process but isolate students so they are less apt to view projects outside of their research area. Locating projects in the halls gives the scientific work more exposure to the students who pass by while changing classes, but the disadvantage is that the projects are spread throughout the school. The facility you reserve should have ample electrical outlets to support those projects requiring electricity.

A third important job of the science fair committee is to establish a *scientific review committee*. According to the *Rules of the 41st International Science and Engineering Fair*, "all ISEF-affiliated science fairs MUST have a Scientific Review Committee composed of a minimum of three persons, including a biomedical scientist (Ph.D., M.D., D.V.M., D.D.S., or D.O.) and a science teacher who is familiar with animal care procedures" (1989e, p. 4). The purpose of this committee is to review each project application that involves vertebrate animals, human subjects, tissue, and recombinant DNA to ensure that the project conforms to the ISEF's guidelines for research in these areas. Establishing this committee may be an administrative headache, but it helps to keep the research involved in fair projects humane and safe. *Appoint individuals on the science fair committee to be in charge of planning the different parts of the fair.* (For details, see box).

THE SCIENCE FAIR COMMITTEE

- *Liaison with the Science Teachers*. This person, probably the chair of the science department, communicates the science fair committee's ideas to the science teachers and brings back their suggestions. The liaison should design the science fair project application form.

- *Secretary*. This science fair committee member should be someone, perhaps a teacher or a parent, with good writing skills. The secretary takes meeting notes and distributes them in a timely manner as a gentle reminder to committee members to get their jobs done. The secretary should also produce typed or word-processed forms, the science fair program, and thank-you letters. If you have a small fair, the secretary might also be able to handle the publicity. This post carries heavy responsibilities, so you might want to make your "secretary" into a subcommittee!

- *Coordinator of Judges*. This person contacts nearby high schools, colleges, universities, businesses, and government agencies to solicit volunteers to help judge the fair. There should be at least two judges for each category in which projects are entered and a ratio of approximately one judge for every four projects. The

(Box continues on page 42)

---

coordinator of judges also designs the judging form. (See below for a sample form.) Science Service publishes an annual *Judging Guide* for its international fair that contains judging information, including explanations of the following ISEF science project evaluation criteria: creative ability 30 points; scientific thought/engineering goals 30 points; thoroughness 15 points; skill 15 points; clarity 10 points.

- *Coordinator of Publicity.* The publicity coordinator puts up posters, distributes fliers, provides items for morning homeroom announcements, and writes releases for the school newspaper and parent-teacher association newsletter. News releases should also be sent to local papers close to the opening of the fair to encourage public awareness and attendance. The coordinator also invites local reporters and photographers to cover both the fair and awards ceremony. If the local press doesn't publicize the fair itself, the publicity person should send the same papers a follow-up story on the winners, the judges, and the scientific exhibits.

- *Awards Coordinator.* Every student who participates in the fair should receive a certificate or ribbon. Such recognition tells students that they are winners for having the imagination and persistence to complete a science fair project. How far the science fair committee goes beyond this minimum recognition depends on the time, money, and energy available. ISEF-affiliated fair directors are eligible to order first-, second-, third-place, and honorable mention science fair medals from Science Service. These directors will also receive special awards from organizations that cooperate with Science Service. The awards coordinator can also ask local merchants, industries, government agencies, colleges, and universities to donate gift certificates, merchandise, facility tours, and arrange meetings with scientists for fair winners.

- *Coordinator of Educational Displays.* If your committee is energetic, it can invite high-technology companies, government agencies, colleges, and universities to set up displays at the fair. A model of the space shuttle or a hydroponic garden will likely fascinate observers. However, some committees choose not to encourage these displays because of the belief that they detract from the prime reason for having a science fair—to highlight the research of students.

JUGGLING SIX BALLS AT A TIME (SUPERVISING STUDENT PROJECTS)

Students who participate in science fairs need guidance and encouragement. Teachers should direct students to sources of information on a research topic, establish deadlines, and help participants when problems arise. The resulting interaction between students and teachers is important. In many cases it generates an excitement about science. See box at end of this paper for tips on designing students' application forms.

Larry Farmer helped Loudoun County's science students (and their teachers) by having a group of teachers write a step-by-step guide for doing a science project. The 27-page handout covers a range of subjects: picking a topic, doing research, forming a hypothesis, deciding on an experimental design, writing the project research paper, developing the abstract, and using the proper bibliographic format. Fairfax County has a similar guide for its teachers and students who work on science fairs.

The teachers at Broad Run High School help their students conduct projects and use time wisely by giving them a "Science Fair Project Time Line" like the one summarized below. Fairfax High School took a slightly different approach, according to Sandy Shockley, science department chairperson (personal communication, Aiello and Hansen, 1989d): Once the date of the local fair is set, each teacher establishes and distributes a science project work schedule for participating students. Both approaches share the purpose of giving structure to the science fair activity and providing teachers with checkpoints to assess students' progress. See below for a sample work schedule.

STUDENT SCHEDULE

- *September 15*. Submit topic.
- *September 29*. Identify articles and books related to the topic and determine which library has them.
- *October 6*. Photocopy 10 articles or relevant portions of books. Include author, title, publication city, publisher, date, page numbers.
- *October 20*. Submit question science fair project will attempt to answer. If the project deals with vertebrates, humans, recombinant DNA, or tissue, submit the appropriate ISEF

(Box continues on page 44)

---

forms at this time. *As mentioned earlier, this is a good precaution even if you are not affiliated with ISEF.*

- *November 17.* Turn in science fair notebook containing the rough draft of the background for the experiment, the hypothesis, and 10 relevant sources (no dictionaries or encyclopedias). At this point, it should be possible to identify the independent variable and the dependent variables.
- *November 22.* Submit procedure (step-by-step explanation of how the experiment will be conducted), a complete list of materials used, a blank data table to show how to keep track of data, and an explanation of how the data will be analyzed.
- *December 15.* Turn in science fair notebook to show that data collection has started.
- *January 12.* Hand in science fair notebook with completed data collection, finished data analysis, completed graphs and/or statistical analysis, and interpretation, which includes an answer to the research question.
- *January 17.* Submit science fair notebook with conclusion and a one-page abstract.
- *February 2.* Turn in final typed paper consisting of title page, abstract, table of contents, purpose question, hypothesis, background information, procedure and list of materials, data, data analysis including graphs and statistical tests, conclusion, and bibliography.
- *February 17.* Bring corrections to final paper, display, science fair notebook, and apparatus to school.
- *February 25.* Science fair.

WHERE TO FIND A TOPIC

There are many books in public libraries on science experiments and projects. Among them are the following:

From Science Service:

- *Guide to the [most recent] International Science and Engineering Fair* lists the 13 categories of the fair, their scientific specialties, and their interpretations. (For details, see below in this section, page 137.)
- *Research of the Finalists* (annual) lists the titles of the projects entered in that year's Fair.

- *Thousands of Science Projects* collects over 7,000 titles of projects completed for the ISEF and the Westinghouse Science Talent Search.
- *Abstracts [most recent] International Science and Engineering Fair* (annual) prints one-page abstracts of the projects at the current year's fair.

From other sources:

- *Science Fair Project Index 1981–1984* (1986), edited by Cynthia Bishop and Deborah Crowe, lists thousands of project titles.
- *Switch on to Science* (1986) (Insights Visual Productions, Inc.), a motivational videocassette, presents ideas for topics. A teachers' manual lists over 400 additional science project ideas.

SET REASONABLE LIMITS

As students choose topics, form hypotheses, and establish research designs, teachers will need to help them be realistic. A student may have a theory about black holes, but if your school doesn't have a million-dollar telescope, there may be no way to examine the hypothesis. Teachers may want to explain to students that parents and scientists can serve as mentors or advisors but must not conduct the research or write the report. NSTA presents guidelines on parental involvement in *Science Fairs and Projects Grades 7–12* (1985), pp. 67–68.

Teachers requiring students to do science projects may want to set aside some time each week for individual conferences. It is this one-on-one contact that Soto of Broad Run High feels is special about science fairs. Soto requires her students to keep science fair notebooks, using bound composition books. In these notebooks, students write the drafts as well as the final versions of each step of their projects. The students are not allowed to add or tear out pages. Every time she checks a step in the time line, Soto stamps the entry with a different rubber stamp. This technique allows her both to follow the development of the students' projects and to make sure that the students are doing their own work.

As science fair day draws closer, teachers may need to help students plan their displays. Vicki Dorsey, a science teacher at Loudoun Valley High School, uses her collection of photographs taken at regional, state, and international science fairs to give students ideas for displays (personal communication, Aiello and Hansen, 1989d). The backboard can be poster board, mat board, plywood, polystyrene, or any other sturdy, lightweight material. Teachers may wish to have students bring their finished displays, project papers, abstracts, and notebooks to school a few days before the fair to check for completeness and embarrassing misspellings.

THE BIG SHOW

Even though the most important learning takes place while the students work on their projects, the most exciting event is the science fair itself. Before the fair, the science fair committee should arrange the display tables, and, if possible, cover them with paper and put paper skirts around them. A local newspaper may be willing to donate the paper. Remember to run extension cords to those tables at which projects requiring electricity are displayed. The committee secretary and helpers should make signs to put up outside the school directing judges and the public to the fair. The secretary should also put up signs in the display area so that students, judges, and visitors can find the different categories.

Use a numbering system to identify the projects. The secretary can make two sets of identically numbered cards, each with a prefix number (1 through 13, if you're following ISEF precedents) indicating the category, and a second number representing the project number within the category (e.g., 6–5, 6–6, 6–7; 11–1, 11–2). Arrange one set of cards on the display tables according to category, and give the second set to entrants as they arrive to set up their projects. The students will know where to put their projects by matching the number on their cards to the numbers on the tables. Prior to the judging, a committee should check that all the exhibits conform to the size and safety guidelines established either by Science Service or by your committee.

Following the judging, which will take several hours, open the fair for public viewing. The secretary should have prepared a program listing project numbers, project titles, and student names. (An index listing all the students alphabetically by their last names is also useful.) In addition, the program can list the judges and give the judging criteria. If appropriate and possible, list special prizes, their donors, and nonstudent exhibitors.

Having the awards ceremony immediately after the public viewing of the project will maximize attendance at both events. If your science fair is held on a weekday, you should consider having an evening ceremony so that working parents will be able to attend. Expect a crowd. Farmer said that there was a larger turnout of parents for Broad Run High School's first science fair awards program than for any other event at the school.

If you have a small fair, present a certificate to each student at the ceremony. If you have a large fair that makes recognition of each student impractical, put the participation certificates at each student's display. Category winners should get a special certificate, ribbon, or medal. The emcee should recognize the donors of special awards as well as the members of the science fair committee, the judges, and any other individuals who assisted with the fair. Students should take their projects home with them following the program since, in many cases, their parents will be available to help.

THE CARE AND FEEDING OF JUDGES

Good judges are important to a science fair, not because they decide who the winners are, but because they help the students continue the learning process that they experienced in working on their science projects. Good judges should be *teachers*. The science fair committee member in charge of judging should arrange for enough judges to cover the fair.

The Loudoun County regional fair and the Fairfax High School local fair each consists of about 200 projects, and each uses 65 to 70 judges. The coordinator of judging should send all judges a letter confirming the date, time, and place of the fair. Along with the letter each judge should receive a sample judging form (see page 52 of this section), an explanation of the judging criteria, and, if the fair is really well organized, copies of the abstracts of the projects the judge will be evaluating. If you are affiliated with ISEF, you may want to create your own pamphlet on judging adapted from Science Service's *Judging Guide* (1989c).

On the morning of the fair, a room should be set aside as a judges' lounge and meeting room. After the judges have had coffee and a small snack and an opportunity to meet each other, give them an overview of the day's events. Be sure to emphasize the role of judge-as-teacher and provide information about the judging process. Before the students arrive, the judges should have an opportunity to preview the projects they will be evaluating so they can develop questions they want to ask the students. When the judges are finished with their preview, the students should be allowed into the display area to talk with the judges. Broad Run provides the judges with three standard questions:

1. What is the purpose of your research?
2. What are your variables and controls?
3. What will be the next logical step in your research?

The judges formulate additional questions based on individual projects. In their scientific dialogue with the students, the judges should be encouraging and should not have unrealistic expectations.

Following the interviews with the students, the judges should be given lunch and then meet by category to select the winners. Although NSTA's judging criteria result in a numerical score for each student, Farmer feels that the judges should use the totals as guides only and reach a consensus about the winners. As an alternative to picking first, second, third, and honorable mention winners, you may want your judges to designate in each category the top three or four winners—without distinguishing among them—who will go on to the next level of competition. The committee should invite judges to the awards ceremony and acknowledge their efforts in the fair.

Be sure each participant receives some kind of award. The science fair experience should encourage all participants.

WINDING DOWN

Is a science fair a lot of work? Yes, but what isn't that is worth doing? Besides turning students on to the magic of science, a science fair has a number of benefits. Teachers benefit by getting to know their students personally and from the emotional recharging they get from their students' enthusiasm. The curriculum benefits because real, experimental science is taking place. Students with a variety of interests interact and benefit: The science student does well in formulating the hypothesis and conducting the experiment; the English student excels in writing the project research paper and abstract; the art student shines in designing and producing the science fair display. In the best projects, these talents intermingle: Some may even be born as the project evolves. The members of your community benefit by becoming more aware of the scientific process and of the importance of science in their lives.

REFERENCES

Aiello, Nancy C., and Hansen, Brian E. (1989a, September 26). Interview with Joy E. Bauserman, science teacher, Broad Run High School.

Aiello, Nancy C., and Hansen, Brian E. (1989b, October 18). Interview with Vicki Dorsey, science teacher, Loudoun Valley High School.

Aiello, Nancy C., and Hansen, Brian E. (1989c, September 19). Interview with Larry Farmer, math and science supervisor, Loudoun County Public School System.

Aiello, Nancy C., and Hansen, Brian E. (1989d, October 16). Interview with Sandy Shockley, science teacher, Fairfax High School.

Aiello, Nancy C., and Hansen, Brian E. (1989e, September 26). Interview with B. Blaine Soto, dean, Broad Run High School.

Bishop, Cynthia, and Crowe, Deborah. (1986). *Science fair project index, 1981–1984.* Metuchen, NJ: Scarecrow Press.

Iritz, Maxine Haren. (1987). *Science fair: Developing a successful and fun project.* Blue Ridge Summit, PA: Tab Books.

NSTA. (1988). *Science fairs and projects grades K–8* (2nd ed.). Washington, DC: Author.

NSTA. (1988). *Science fairs and projects grades 7–12* (2nd ed.). Washington, DC: Author.

Science Service. (1989a). *Fortieth International Science and Engineering Fair.* Pittsburgh: Author. (Abstracts)

Science Service. (1989b). *Fortieth International Science and Engineering Fair.* Pittsburgh: Author. (Guide)

Science Service. (1989c). *Fortieth International Science and Engineering Fair.* Pittsburgh: Author. (Judging guide)

Science Service. (1989d). *Fortieth International Science and Engineering Fair.* Pittsburgh: Author. (Research of finalists)

Science Service. (1989e). *Fortieth International Science and Engineering Fair.* Pittsburgh: Author. (Rules)

Switch on to science. (1986). Encinitas, CA: Insights Visual Productions. (Videocassettes, 35 minutes)

Yoshioka, Ruby. (Ed.). (1987). *Thousands of science projects* (2nd ed). Washington, DC: Science Service.

TIPS ON DESIGNING THE SCIENCE FAIR PROJECT APPLICATION FORM FOR INDIVIDUAL STUDENTS

- Besides providing obvious information (name, date, teacher, grade, project title, project description), the students should indicate the research category to help anticipate the size and the number of judges needed in each category. The ISEF recognizes 13 categories. (For details, see page 137.)
- Find out if the final display requires electricity to plan for outlets and extension cords.
- Ask if a computer will be a part of the display. If so, supply a surge suppressor or require the student to do so.
- Determine if the project involves vertebrate animals, human subjects, recombinant DNA, or tissue. If it does, participants must comply with the special rules outlined in *Rules of the [most recent year] International Science and Engineering Fair*. This pamphlet also contains permission forms for experiments in each of these areas. Fairfax High School staples the four sets of ISEF guidelines and forms in packets with different colored covers. It's probably a good idea to follow ISEF rules in this sensitive area even if your fair is not an affiliate.
- Specify the maximum size of students' final displays. ISEF limits are 76 cm deep, 91 cm wide, and 2 m, 70 cm high (floor to top).
- Furnish other necessary information on the application: application deadline date, date of the fair, list of other required documents, such as research paper with a bibliography, one-page abstract, and project record book.
- Require signatures of student, parent, and supervising teacher.
- See NSTA sample form below in this section, page 51.

Nancy C. Aiello, chair of the science division at the Loudoun campus of Northern Virginia Community College, has been the director of the Virginia State Science and Engineering Fair since 1985. Brian E. Hansen, who won third place in the Napa (California) High School science fair in 1963, is now an associate professor of English at the Loudoun campus in Sterling, Virginia.

Science Fair Project Application

Name _____ Date _____

Teacher _____ Grade _____

Project Title _____

Project Description (be brief) _____

_____

_____

PROJECT AREA (circle one):

Biology   Chemistry   Physics   Mathematics   Behavioral   General Science

PROJECT TYPE (check one):

_____ *Experimental*—Forming a hypothesis (question) about something the student doesn't know the answer to, doing an actual scientific experiment, making observations, collecting data, and reaching conclusions.

_____ *Demonstration*—Science in a show and tell format. The student knows what is going to happen when he or she begins. Includes models, kits, collections, posters, etc.

_____ *Biological*—A project involving living things such as insects, birds, food, people, diseases, etc.

_____ *Physical*—A project involving things not living such as chemicals, stars, air pressure, weather, etc.

Will you require electricity? _____ YES _____ NO

Your project should include the following items:

    1. Exhibit that can stand by itself.
    2. Research paper with bibliography.
    3. Abstract (one page summary, with bibliography).
    4. Materials necessary for the exhibit.
    5. Oral presentation (3 to 5 minutes).
    6. Logbook of daily work.

Return this completed form to your teacher by _____

Student signature _____

Parent's signature _____

Teacher's signature _____

*Reproduced with permission from *Science Fairs and Projects, Grades 7-12.* Copyright © 1988 by NSTA.

SCIENCE FAIR PROJECT JUDGING CRITERIA*

Scientific Thought (30 points)

Does the project follow the scientific method? (hypothesis, method, data, conclusion)
Is the problem clearly and concisely stated?
Are the procedures appropriate, organized, and thorough?
Is the information collected accurate and complete?
Does the study illustrate a controlled experiment that makes appropriate comparisons?
Are the variables clearly defined?
Are the conclusions accurate and based upon the results?
Does the project show the child is familiar with the topic?
Does the project represent real study and effort?

Creative Ability (30 points)

How unique is the project?
Does the exhibit show original thinking or a unique method or approach?
Is it significant and unusual for the age of the student?
Does the project demonstrate ideas arrived by the child?

Understanding (10 points)

Does it explain what the student learned about the topic?
Did the student use appropriate literature for research?
Is a list of references or bibliography available?
In the exhibit, did the student tell a complete and concise story, and answer
some questions about the topic?

Clarity (10 points)

Did the student clearly communicate the nature of the problem, how the problem
was solved, and the conclusions?
Are the problems, procedures, data, and conclusions presented clearly, and in
a logical order?
Did the student clearly and accurately articulate in writing what was accomplished?
Is the objective of the project likely to be understood by one not trained in the subject area?

Dramatic Value (10 points)

How well did the student design and construct the exhibit?
Are all of the components of the project done well? (exhibit, paper, abstract, log of work)
Is the proper emphasis given to important ideas?
Is the display visually appealing?
Is attention sustained by the project and focused on the objective?

Technical Skill (10 points)

Was the majority of the work done by the student, and was it done at home or in school?
Does the project show effort and good craftsmanship by the student?
Has the student acknowledged help received from others?
Does the written material show attention to grammar and to spelling?
Is the project physically sound and durably constructed? Will it stand normal wear and tear
Does the project stand by itself?

*Reproduced with permission from *Science Fairs and Projects, Grades 7-12*. Copyright ©
1988 by NSTA.

PRESENTING SCIENCE PROJECT RESULTS: VISUALLY (THROUGH POSTER PROJECTS) AND ORALLY (THROUGH SCIENCE SYMPOSIA)

John A. Blakeman

POSTER PROJECTS

Scientists present their findings through a number of media, including oral presentations, journal articles, and increasingly at scientific meetings and symposia, by the use of posters. Like a good science fair display, upon which the poster project is based, a properly prepared poster conveys a large amount of information in a small space, in a short period of time, and often in the absence of the author. Robert A. Day, author of *How to Publish a Scientific Paper* (Third Edition, 1989), explains this method of presentation as being part of the "trickle-up" effect, as scientists who were also parents helped their children with science fair projects and thought up the poster projects as an alternative to the formal papers that were traditionally delivered orally. The poster is a useful information medium with which the literate science student should be familiar.

In planning and constructing the poster, follow the guidelines below. Especially note the comments and reasons for each item.

ELEMENTS OF THE POSTER

The poster should include these elements:

- title
- topic headings
- topic heading captions
- topic texts
- reference section
- figures
- exhibits (optional)

TITLE. The title, often effective in the form of a question, should quickly explain or introduce the subject of the poster. Lengthy, sentence-like titles should be shortened to be easily read and fitted within the title area of the poster display. The title should have the largest lettering of any of the elements but not be overbearing.

If the display board has a center section with two adjacent wings, the title must appear either within the upper one-third of the center section, or in the upper one-fifth of both the wings and center sections. Other display arrangements, such as those made up of a single board, should place the title similarly. The title is important, but should not consume a major portion of the poster.

TOPIC HEADINGS. The poster should have several major topic headings, using letters much smaller than the title but quickly readable in a brief scan of the display. Topic headings can be single words or short phrases. They introduce or note the major topics of the subject, and should appear in an orderly, sensible arrangement, and in parallel grammatical forms.

TOPIC HEADING CAPTIONS. Under each of the topic headings should be a sentence or short paragraph that briefly and simply explains the topic. These words, in smaller print than that of the headings, should be in sentences (not just terms or phrases). In many cases, printing the topic heading captions in capital block letters is effective. Any variation that accomplishes the purpose of informing the viewer quickly about the project's major details is acceptable. The topic heading captions should convey quickly the major points of each topic to a disinterested viewer.

TOPIC TEXTS. Beneath each topic heading caption (the major points of each topic), place the topic text, a typed or neatly handwritten explanation of the heading. Material is presented in detail—usually more than is wanted by the casual viewer. Only those who are really interested in your topic are expected to read the text; therefore, everything else in your display must give enough information to make your poster interesting when the text goes unread—it mostly does. The quality and depth of the work on your subject will be most evident in the topic texts.

Just as in a scientific paper, topic texts should contain proper citations to the sources of your data. Consequently, the poster should have somewhere a labeled references section. Citations appearing in all topic text sections should refer back to a single reference list on the poster. This list should be small in comparison to the other elements of the poster and appear in a nonprominent corner. The reference list offers evidence that you have researched your subject in a scholarly manner.

FIGURES. A figure is a picture, diagram, drawing, or other illustration. Almost every project is improved by graphic illustrations, especially if the viewer is unfamiliar with your subject. Each figure must have a clear, informative title and caption. The essential nature and application of each figure should be evident without reference to the text: that is, the caption should be complete and self-explanatory.

Figures should not be too big. Most should not be larger than a standard sheet of paper.

EXHIBITS. Many poster displays are effective in part because they include specimens, apparatus, or other three-dimensional items. Exhibits are optional (and inappropriate for many subjects). When exhibits are used, be sure to label and explain each important item.

PLANNING THE POSTER

Construct your poster following a plan. This sequence may help:

1. Make a list of the major topics.
2. Assign short, descriptive titles to the poster and its topics.
3. Arrange them in an orderly sequence.
4. Compose preliminary captions for them.
5. Write preliminary texts for each topic heading, including the reference section.
6. Choose and construct the figures.
7. Make a small-scale layout or sketch of the poster.
8. Print topic headings, captions, texts, and figures on separate pieces of paper that will be attached to the poster.
9. Print the major title either on the poster or on paper to be attached to the poster.
10. Assemble the poster by attaching all the separate papers.

GENERAL HINTS

Do not attach printed folders, pamphlets, or other literature. These may be cited in the text, but should not physically appear on the poster. Avoid photocopies, except for properly attributed figures. Freely use color, but avoid ribbons, bows, strings, or other purely decorative items that do not contribute directly to the meaning of the display. Do not attach portions of any research paper you've done.

ELEMENTS OF THE SCIENCE SYMPOSIUM ORAL PRESENTATION

I. State the setting of the problem [BACKGROUND INFORMATION]. Be specific to your system. Presume that your listener is scientifically and mathematically literate but is not familiar with the details or setting of your problem. When appropriate (check with advisor) try to mention at least two authors or authorities whose work underlies your project. This lets the listener know that you have referenced your work. The purpose is not to impress the listener with your knowledge, but to demonstrate that you have done a good literature search.

II. State the question(s) you are attempting to answer [THE PROBLEM]. Begin this section by stating the title, "*The Problem.* The problem this research project investigated..." Draw the listeners' attention to this important statement by a brief pause, so that they know that you are finished with background information.

III. State the hypothesis of the research [THE HYPOTHESIS]. As above, begin, "*The Hypothesis.* My hypothesis is that..." Offer a brief, well-considered statement about what you expected to happen or to be proved by your work.

IV. Theoretical basis of the research [THEORY BEHIND THE HYPOTHESIS]. Immediately after the hypothesis, tell why you thought your hypothesis was appropriate, referring to previous work or information pointing toward your hypothesis. Cite other researchers, when relevant to your hypothesis. A good beginning to this section would be, "This hypothesis is based on..."

V. Methods and materials [HOW THE PROJECT WAS DONE]. Begin by stating, "Methods and Materials. Research began by..." Tell enough that a listener (with a written copy of your complete materials and methods section) could accurately *replicate* your work. Be appropriately detailed. Practice this part of your presentation in advance.

Visual aids can be helpful. If an overhead projector is available, transparencies may help illustrate major procedures. In poster sessions, point to figures on the poster. Include three-dimensional displays or apparatus if appropriate.

VI. Results [WHAT YOU FOUND]. Begin this section by a statement such as, "I determined that..." Present the results of your project in logical (*not necessarily chronological*) order.

Use some sort of visual aid here—either a transparency, poster, graph, table, or other representation of your results, often in quantified form. Your audience should both hear and see the results.

VII. Confirmation or negation of the hypothesis [WHAT WAS PROVED]. In this section, first restate the hypothesis and then continue that "the hypothesis was confirmed (or denied)." Explain how your research data supported or undermined your thesis, *pointing* to specific findings in your Results figures.

VIII. Discussion [IMPLICATIONS OF THE RESEARCH]. Continue smoothly from the last section. The topic will alert the listener. In this last section, talk about subjects that don't fit elsewhere, focusing especially on relevant personal or subjective observations, hunches, opinions, or comments. *So label such speculations.* State the practical or scientific implications of your research. One effective final statement suggests that your findings just presented provide the basis for further research.

For John A. Blakeman's biography, see page 6.

HOLDING AN INVENTION FAIR
Donald R. Daugs

The recent trend of incorporating science–technology–society concepts into the 7th–12th grade curriculum has helped promote invention fairs. But, like science fairs, invention fairs take time and do not always fit into the day-to-day curriculum. If successful, however, invention fairs are a stimulating club activity.

Unfortunately, inventiveness and creativity are not always regular fare in science programs. Even less evident is an emphasis on technology. Outlined here are steps that will spark creativity and engage technological process skills. The logic in the following process may never consciously occur to club members, but if the participants follow the steps, students may produce some inventions. The process seems to work as well with 12 year olds as it does with adults. Teachers and students will find enough detail in each step to give the club leader ideas about what to do, but the leader needs to be creative in fleshing out the process.

Step 1 (one hour). Introduce the concept of inventions by putting a string, a button, a bean, and a paper clip on the table. Challenge club members to make something useful from the objects. This experience should involve play and exploration. Share the products, then introduce the idea of an invention fair. Indicate that the next few club sessions will focus on improving ability to make inventions.

Step 2 (one hour). Devote a session to "brainstorming," a group method for generating a large number of ideas. Keep these points in mind:

• List all ideas.
• Accept everything. Don't be critical or evaluate ideas generated.
• Encourage the unusual.
• Don't stop too soon. Incubation (thinking) time is important.

The club leader should be a facilitator, catalyst, and a cautious contributor in the process.

Sample Activity:
   Nearly everyone likes ice cream. Brainstorm ways you could bring ice cream from home for lunch at school.

Step 3 (one hour). Devote part of a session to the technique of building alternatives. Such an approach encourages examining things from various perspectives, and promotes restructuring or rearranging of information. Phrases that encourage building alternative thinking include

- "How many uses can you think of for…?"
- "How many ways could you tell someone…?"
- "Think of different ways to…?"

Sample Activity:
    Display a steam iron. Challenge club members to think of different uses for the iron. Have students work either as a total group or in smaller groups of four to five members.

Step 4 (one-half hour). Devote part of one session to the skill of synthesizing or of putting things together. Creative synthesizing combines things in new or different ways to produce a new product.

Sample Activity:
    Have students think of what they could add to a science textbook to make it more useful.

Step 5 (one hour, with possible extension). Devote one session to the skill of changing parts. One way to devise a new invention is to change one or more parts of an existing object or process. This skill starts with describing the attributes of an object. Attributes of noodles include color, size, shape, and flavor. Numerous factors influence each of these attributes. For example, the amount of salt influences flavor but probably has no effect on color, shape, or size.

Sample Activity:
    Using the following recipe, make no-bake cookies.
        250 ml honey
        250 ml peanut butter
        375 ml powdered milk
    Blend the ingredients together and shape into small balls. Roll the balls in coconut shreds or powdered sugar.
        List the attributes of both the ingredients and the end product. Speculate on what the product would be like if proportions of the original ingredients, or the order of combining ingredients were changed. Make some changes with smaller batches. See what happens. Add new ingredients. Invent a better no-bake cookie. When you have something great, name it, make a large quantity, and have a cookie sale to raise money.

Step 6 (1 hour). Devote one session both to visualizing—the process of generating mental images—and to "think drawings"—putting the ideas into pictures. Often, you can draw a mental image more easily than you can describe it. "Think drawings" should illustrate ideas, functions, or actions, not texture, color, or quality.

Sample Activity:
Visualize a car wash, a shower, and a bath tub. Mentally put a person and a car in each. Now invent a people washer that utilizes the best features of a car wash, a shower, and a tub. Make a drawing of your invention. Label the parts and explain how they work.

Step 7 (lots of time). Challenge the club members to invent a new product. Incubation of ideas as well as utilizing the skills described in Steps 1–6 are important. The motto, "No one of us is as smart as all of us," may apply. Club members should work in groups. Leaders or parents may need to discuss alternatives. The end product of this step should be a detailed plan that is the beginning of a diary or log. Any notebook will work. Comments, correction, and reviewers' initials may be placed in the left margin. Eventually this log may include

- how the person got the idea for the invention
- background information
- steps followed in developing the idea
- problems encountered and how they were solved
- drawings and procedures
- test results
- photos
- names of people consulted and what they contributed

This procedure is the best protection if you file a patent application.

Step 8 (variable). This is the "do it" step. Physically create what you have been mentally planning. In industry this stage would be the development and testing of the product to evaluate prototypes. Evaluation considerations may include

Need
1. Will someone use the invention?
2. What are the invention's advantages and disadvantages?
3. What are the potential impacts of the invention?

Practicality

1. When, where, and how will the invention be used?
2. Is it complex or simple and easy to use?
3. Is it easily damaged?
4. Is it cost effective?

Originality

1. Is there a similar product already available?
2. Is it creative or unusual?
3. Is it cleverly named?
4. Can it be recycled?

Step 9 (2 hours). One major session should be devoted to an invention fair. Club members should display and demonstrate their products. If they are going to be judged, the following criteria may be of value. The invention

• reflects original, creative thought
• has practical value
• is carefully designed and constructed

The inventor

• has provided evidence to show that no similar product or process exists
• is enthusiastic about the invention
• has made wise use of available materials to control costs
• has promoted the product effectively

The above items, if each is rated on a 5-point scale, total a possible 35 points for judging.

One final step: You may want to contact local business to offer prizes for your inventions.

REFERENCES

Daugs, Donald R. and Monson, Jay A. (1989). *Science, technology, and society: A primer for elementary teachers.* Logan, UT: Utah State University.
Hindle, Brooke. (1983, February). Human nature and the contriving mind. *Science Digest,* 91(2), 48.

Donald R. Daugs, professor of elementary education, teaches science, technology, and society concepts at Utah State University. He conducts club activities at the Edith Bowen Laboratory School for faculty and students.

DOING A SCIENCE RESEARCH PROJECT

Gene Kutscher

Twenty-three local high schools on Long Island, and dozens of others—public, private, and specialized—around the nation have started independent science research clubs or programs. These programs show, once again, that there is no better way of learning science than doing science. And, although friendly rivalry exists among these programs, the teachers share ideas and assist other schools to start up independent programs in a noncompetitive, collegial manner.

Independent study programs have an important advantage over traditional science fairs because of the lack of time constraints. Because of this freedom, concepts can be investigated in depth. Independent study typically lasts from several months to four years. Different schools use varied approaches to independent study, but all participants share the philosophy that the true task of a scientist is to do research.

In our nation, we expect professional athletes to develop by practicing intensely from as early as their middle school years. In contrast, we expect science researchers to blossom much later, at the university level. Such expectations are illogical, and in all likelihood, we lose scientists because of them. Independent science research programs may help correct this problem. Young people are capable of, and, in fact, do excellent and original research at the high school level. All students who are either academically talented in science or who simply love to ask "Why?" should be encouraged to investigate original ideas. Canadian students are already doing independent research. In British Columbia, for example, the Canadians have started a pilot program with the ultimate goal of developing a national program in student science research.

What are the advantages of independent science research for the student? For one thing, the flexible time frame allows experiments the time needed to evolve. For another, the students are able to investigate an idea that interests them. In addition, they develop a keen understanding of what science research entails, ranging from the frustrations of long hours with inconclusive results to the intense joy of realizing that they have discovered something new. Of course, along the way they develop excellent skills in problem solving in a manner far more lasting than that gained by writing and memorizing the steps of "the scientific method." A side benefit, which, of course, should not be a *primary* reason for students to conduct independent study, is that colleges often like applicants who have done such work.

GETTING STARTED

To learn from experience how to start an independent science program, take a look at how one successful, average-size, public high school did so. The problems the program encountered are included as well as how they were overcome. Then, we'll turn to other varieties of programs.

At Roslyn High School on Long Island in New York, teacher David Berry started a high school research program in 1972. At that time, one or two students a year had asked to conduct independent study. Berry arranged to assist them in lieu of a class assignment whenever the students and teacher were able to meet: before or after school, during a lunch period, or during mutual free time. Berry's method of meeting is still in place at Roslyn today; however, the program has expanded to 80 students (or about 10 percent of the school), and many advisors participate.

When Berry left in 1976, Roslyn hired Charles D. Duggan to run the program. Duggan had previously developed and successfully directed a large science research program in St. Louis. In his first year at Roslyn, he served as mentor to seven students in research in lieu of a class assignment. During Duggan's second year, because of his enthusiasm and encouragement, 14 students joined the program. Duggan asked to help guide these students instead of teaching two classes; the district wanted his teaching load reduced by just one section. Both points of view had certain merits. Few districts can afford two-fifths of a teacher's time to support 14 students. However, arranging meetings and assisting more than 10 researchers meaningfully at one time is difficult. Eventually, only seven students were placed in one section. The other seven highly interested students could not participate.

By 1980–1981, the basic structure of the Roslyn program was in place. It began with "Introduction to Research," a semester-long, one-half credit course, which is required for all students in the program. To allow students time to adjust to high school, Roslyn does not allow them to take the course before the second semester of ninth grade; however, they may not enroll later than the first semester of their junior year. Meeting five periods a week in a regular class setting, the course is graded high–pass, pass, or no credit. The goal of "Introduction to Research" is to develop a workable project proposal. Once the student selects a project, s/he is assigned to a teacher for guidance. The student receives a half credit per semester of work, evaluated in the same way as "Introduction to Research." Often, an additional volunteer specialist in the field also is mentor to the student.

During 1980–1981, three to five students were assigned to a research advisor in lieu of two and a half hours of formal teaching each week. (A full teaching load is 25 hours per week.) As the students started to enter competitions, such as the prestigious Westinghouse Science Talent Search, and win awards, the program became popular with both students and their parents. There were 14 students in the program in 1978–1979, 28 in 1979–1980, and 40 in 1980–1981. In order to accommodate diverse student interests in research, many additional teachers became advisors. This year (1990), 64 students are conducting research, more than 25 are taking "Introduction to Research," and 6 teachers advise independent study.

The program does not benefit merely the students. Both *The New York Times* and *Newsday* (Long Island's largest circulation newspaper) have cited the science research program in their real estate sections as reasons to move to Roslyn, making good public relations for school and community alike.

OTHER OPTIONS

Other schools have differently organized research programs. A few schools have research clubs with paid advisors. This arrangement has an advantage in accommodating small numbers and providing high excitement. But it also has a disadvantage: Meetings occur after a long and full school day.

Several schools have made research a part of the regular honors program. In such schools, where honors classes typically met for double and single periods on alternate days, honors classes increased to double periods every day, with a half a period a day devoted to research. This method offers the plus of a set time and place to meet during the regular school day and a relatively low cost (25 students per 2.5 contact hours). However, it has several strong minuses as well, including the limitation of available research to honors students, the requirement that all honors students conduct research, and the teacher's difficulty of working at once with many projects reflecting diverse student investigations. (Science teachers are knowledgeable, but few can stay at the frontiers of all of the sciences!)

Still another method of organizing an independent research program is reflected by one school that sends most of its population to local hospitals and laboratories to work. This approach is low in cost and offers students outstanding experiences working with complex equipment at ongoing projects. If this path is pursued, educators must insure that the students are working rather than washing glassware and that they are understanding what they are doing. For example, the Bronx High School of Science, a selective specialized school, which annually admits (on the basis of scores on a competitive examination) approximately a thousand students, maintains files of adults willing to act as mentors. It also keeps model letters of introduction, expectation, and thanks. Students are briefed about what to expect and how to act. Teachers follow up by meeting regularly with the students and by sending their own thank-you notes after the work is completed.

The main disadvantage in this type of program is that students often must pursue whatever is happening in the particular laboratory to which they are assigned. A lesser disadvantage, but one that is important to purists, is that many times students do not completely follow the process of investigating, obtaining, and setting up equipment. They must arrange, moreover, for transportation to and from the lab after school.

Other schools concentrate on summer work for the students in local hospitals, laboratories, and universities. These programs are similar to the one described above, but better, because students can devote full time to research. Marilyn Hanson, an NSTA Presidential Award winner at Madison Memorial High School in Wisconsin, uses this method. If they wish, students can make arrangements to continue the research during the school months. (See box on next page.)

In summary, many possibilities exist for starting a research program. Schools that continue to be cited for excellence in science, such as Roslyn and the Bronx High School of Science, pursue combinations of in-house, at-home, university, and laboratory work, as is most appropriate to the specific project and the individual student.

THE TIP OF THE ICEBERG: WHERE TO LOOK FOR MORE SCIENCE AND MATH EVENTS

There are a considerable number of summer opportunities for science and math study and competitions not covered here because they change so rapidly and because they are *mostly* best investigated locally rather than through a guide covering such events nationwide. Science Service, however, helpfully lists many such opportunities in its *1989 Directory of Student Science Training Programs*, its 612 entries comprising 89,176 openings. Students should take care when choosing a program to insure that it is among the few—only a dozen or so—that are research-immersion programs rather than course-oriented ones. Another publication, *A Taste of College: On-Campus Summer Programs for High School Students*, by Jane E. Nowitz and David A. Nowitz notes university and college classes open to some high school students (7 in the natural sciences, 76 in computers and computer science, and 85 in math).

In addition many math and science enterprises occurring both in the summer and in the regular academic year are, though extremely valuable, entirely unofficial. Industries may offer internships; labs may have positions that turn into mentor/apprentice situations; scientists may be willing to trade some of their expertise for assistance. The particulars of such opportunities change rapidly; thus, enterprising individuals—students, teachers, parents, and the like—must seek such openings on their own. State Science Teachers Associations or local groups or individuals such as science supervisors will help.

Readers may find further assistance in three publications covering some materials similar to those in this book. Judith M. Freed has prepared a guide to student competitions and publishing, now in its third edition. She plans a new edition in 1990. Melanie Krieger is at work on a book called *Entering Science Research Competitions*, slated for publication in 1991. The National Council of Teachers of Mathematics intends to update its 1982 booklet *Mathematics Contests*.

WHICH STUDENTS SHOULD PARTICIPATE?

Students who are academically talented in science or those who are curious and interested should be encouraged to pursue independent study. In Roslyn, teachers of the accelerated eighth grade and honors ninth grade classes (Earth science) look for students who probe and are incisive. The teachers ask open-ended questions. Students work on assigned honors projects, investigating anything (the teacher approves) about Earth science that interests them. The best achievers are encouraged to select "Introduction to Research" as an additional elective to their regular science program. Students in other classes who display interest and determination in a particular area are encouraged, or ask, to enter the program as well.

In other schools, admission to the program may be by competitive examination, by academic achievement, or by admission to an honors program. But, open admission to all students who care about science is the most democratic and the most productive policy.

"INTRODUCTION TO RESEARCH": CHOOSING A PROJECT

The goal of "Introduction to Research" is to develop the skills and procedures by which students may focus their interests, develop a proposal, and conduct original research. Students read Peter B. Medawar's *Advice to a Young Scientist* (1979). The course covers

1. What is research?

    • Why conduct it?
    • What are its ethical, financial, and practical restrictions?
    • What are the characteristics of good and poor research?
    • What is "the scientific method"?

2. How does one do preliminary research and select a topic?

    • sources of information
    • library work
    • interlibrary resources
    • brainstorming
    • the problem statement
    • project costs, safety, and testability

3. How does one develop research skills?

   - photography
   - animal maintenance
   - using computers
   - analyzing data, including statistics and graphics
   - using specific equipment

4. How does one present findings?

   - writing a scientific paper
   - preparing for an oral presentation

Students are encouraged to investigate any topic that interests them, as long as it is safe and not too expensive. Students have done research in archaeology, astronomy, biology, chemistry, Earth science, ecology, energy, mathematics, physics, psychology, and sociology. Topics in fields like computer theory and robotics are possible, too.

SCHEDULING

A brief word about scheduling the program is in order, to maximize opportunities and minimize errors. First, to allow interested students to transfer, "Introduction to Research" should not be held in the same time period as the honors science course(s) in the grade from which it will draw the most students. Second, administrators assigning teachers to the program and students to teachers, should honor individual interests. At Roslyn, all but 2 of the 15 science teachers (counting the chair, who coordinates the program) have acted as teacher advisors to individual students at one time or another. For example, eight of the teachers have advised students on Westinghouse papers. This practice gives the department a broad base of knowledge about the program. Several of the teachers like research more than do others, however, and an effort is made to assign these to the program to work with students interested in their specialty. Happily, all of the teachers are more than willing to help out with ideas and suggestions. Similarly, a student with an interest in marine biology research should be assigned to a teacher with experience and expertise in that field.

REQUIREMENTS AND HOPES

Students conducting research should meet with their advising teachers at least once every two weeks from a few minutes to many hours, depending on the stage of the work. Students keep a log and write a report on their progress at the end of each semester.

Teacher–student meetings must be at mutually agreed-upon times. On the one hand, if students miss a meeting without informing the teacher, they are charged with a "cut"; on the other, teachers must also meet with students regularly, in order to keep projects progressing.

At Roslyn, in a program called "Sixteen Wednesdays of Science Research," 16 students present the results of their research to their peers at student seminars. Students even present projects which did not fulfill their original hypothesis. The students learn that all results are important and that there is joy of discovery—even when the evidence denies the practicality of the approach investigated. Thus, they learn the importance of intellectual honesty. The talks are attended, at this open-campus school, by friends and supporters from other classes, by other research students, and by entire classes with an interest in the topic. Attendance ranges from about 10 to as many as 140, with an average of about 75. To attend a talk, students must obtain permission from their teacher to leave class. The talks are spread throughout the day so that no period loses students excessively. While talks are not required, students are encouraged to present their work. For many, the research presentation is their public speaking debut, and they rehearse in preparation. This kind of presentation offers opportunities for important growth.

A similar, but required, program called "An Evening of Science Research" is presented for the community every spring. Several students give brief talks about their studies, and then the work of the entire program is displayed in the cafeteria. This evening is the "research fair" version of a science fair.

Students are encouraged, but not required, to enter some of the many competitions available to them. It is wrong to force students to compete if they are going to develop strong anxieties; however, students unaware of the value of their work and of the universality of nervousness when competing should be coaxed into participating. Even if students "lose," the experience can be valuable. The development of presentation techniques, whether oral, written, or graphic, enhances the total education and confidence of these students. Opportunities are present, as well, for interdepartmental cooperation in the construction and presentation of models, displays, and, occasionally, the whole project. Industrial arts and art teachers, for example, are often wonderfully cooperative and creative.

Among the competitions that students enter are the Westinghouse Science Talent Search, the Space Shuttle Project, the Duracell NSTA Competition, the local versions of the Junior Science and Humanities

Symposia, the International Science and Engineering Fair, the Science Congress, and the Junior Academy of Science. Roslyn students compete in many state and local competitions, too: New York has a Student Energy Research Competition, for example. Through this program, utilities companies, the State Education Department, and the Science Teachers Association of New York State have joined to award over 100 competitive grants for high school level research into energy conservation and/or alternatives. Students may work alone or with as many as two others. If they win a grant (up to $500), they are brought to Albany for a three-day second round competition at an energy fair. Other states have emulated this program. Teachers can petition utilities to develop similar programs elsewhere. Energy research makes ecological sense and is the kind of research that can be approached by any level of student with an interest.

COSTS AND FACILITIES

The greatest cost for high school science research programs is for staffing. One way to build support for starting up is to have administrators and community members visit a nearby program. Roslyn has had more than 50 visits by administrators, teachers, and students over the past few years, and their favorable impressions have resulted in numerous new programs. Most of the schools conducting research across the nation welcome visitors and share ideas. Teachers generally recognize the excellence of research as a vehicle for student comprehension of the processes of science and for its need in all schools. (Some community members unfortunately question why there are no Westinghouse winners at the local school; such a beginning often results in a program whose goal is a prize rather than good science education.)

At Roslyn, the program is allocated about $55 per pupil per year for supplies. Though estimating is difficult, this appears to be about the right amount: Many projects cost more, and many cost close to nothing. The energy competition has relieved some financial pressure by providing funding for several students each year. Equipment is occasionally purchased if it can later serve in the regular classrooms. In general, equipment is borrowed, donated, or obtained on grants. Much of it recycles from project to project.

Projects done at school require space. With enrollment declining in many areas of the country, this room may be available. A research room and a storage area with controlled access are useful.

SOME ADVICE

Get help. Teachers from Roslyn and many other schools present workshops on starting programs at NSTA regional and national conferences.

Call, write, visit, or have research teachers visit you. Remember:

- Start small.
- Obtain funding.
- Build support.
- Maintain public relations.
- And step towards scientific excellence.

See page 14 for Gene Kutscher's biography.

REFERENCES

Freed, Judith M. (Ed.). (1988). *Freed's guide to student competitions and publishing* (3rd ed.). Delaware, OH: Author.

Johnson, David R., and Margenau, James R. (1982). *Mathematics contests*. Washington, DC: National Council of Teachers of Mathematics.

Nowitz, Jane E., and Nowitz, David A. *A taste of college: On-campus summer programs for high school students*. (1989). New York: Simon and Schuster (Arco).

Science Service. (1989). *1989 Directory of student science training programs*. Washington, DC: Author.

Science Competitions—
Nationwide and Worldwide

SCIENCE COMPETITIONS: AN OVERVIEW

Arthur Eisenkraft

On a March day in 1986, a ninth-grade student from Grand Rapids, a small town in Minnesota, accompanied by his parents and his electronics teacher, arrived at the Waldorf–Astoria Hotel in New York City. The trip and the city were exciting for the boy and his family, who had never before flown on an airplane. The next day, he was to receive a $10,000 college scholarship. Little did he or anyone else guess that the events of this day would also change his life.

WEIGHING BIRDS AND LIGHTING SNEAKERS

The student, Melvin Holmquist, had built a device that weighed birds. Melvin enjoyed watching birds from the window of his house. He realized that, although his reference books occasionally gave approximate bird weights, there was little discussion of the range of weights within a species or of the change in weight of an individual bird during a season. Melvin's bird-weighing device was his winning entry in the Duracell NSTA Science Scholarship Contest.

At a press conference in the Waldorf–Astoria, Melvin described the operation of his device:

- A bird rests on a perch at a feeding station visible from Melvin's window.
- The bird's weight causes the perch to slowly descend, triggering an alarm inside the house.
- Hearing the alarm, Melvin goes to the window, records the bird's species, and turns on the meter in his house.
- As the perch descends, a small opaque screen in an enclosed box beneath the perch rises, thereby exposing a set of photocells and simultaneously lighting a bulb.
- The more the perch descends, the more photocells are exposed.
- The amount of light hitting the exposed photocells sends an electric current to Melvin's calibrated meter in the house.
- Melvin records the bird's weight.

Melvin returned home with a scholarship and a new sense of confidence. Three years after the ceremony in New York, I learned that Melvin had become quite popular in the high school upon his return, even something of a celebrity. He also scrapped his plans for taking a job right after high school and directed his sights on a degree in electrical engineering. This scholarship program had a powerful outcome.

RUBE GOLDBERG LIVES

After Melvin had explained how his device worked, a second-place winner from New Jersey was anxious to describe his project— lighted tennis shoes. This boy had inserted pressure pads into the soles of his sneakers. Along the edges of the sneakers were little red and green lights. As he walked the pressure pads would complete circuits and turn the bulbs on and off in different patterns. He said they looked "great" when he walked or when he danced.

Part of my job as a judge for the competition requires me to call students who create the top entries a few weeks before the prizes are awarded. When I called this boy and inquired how he got the idea for lighted sneakers, he explained that, having just worn out a pair of sneakers, he decided to see how they were made before throwing them out. He took one sneaker to the basement, sawed it in half, and ran upstairs to his mother to show her all the unused space in the sole of the sneaker. He exclaimed, "Do you know how much electronics you could put down there?" As a judge and as a high school teacher, my immediate thought was that this was just the type of inquisitive student we would want to reward and encourage with a $3,000 prize.

There were other winners that day in New York: There have been 41 every year since this competition began in 1983. There was a girl who invented a way to view the night sky constellations and star charts simultaneously. There was a student who worked in a nursing home and invented a machine to deal playing cards to help arthritic patients enjoy themselves. There have been lighted dog collars, alarms, electronic rulers, and protractors. Students have found new ways to feed fish, to protect their bicycles, to dispense toothpaste, and to measure wind speed. Duracell sets the challenge, teachers notify and encourage students, and the students use their imaginations to conceive ideas and their skills to turn these ideas into working devices.

BATTERIES AND BEYOND

Not everyone in the United States interested in science was worrying about batteries last year. One high school student was exploring ways in which E. coli could be modified during a gene-splicing experiment. Another was trying to find similarities between the mathematical structure of the equations describing electromagnetism and those describing general relativity. And still another was recording life cycles of a snail exposed to different environmental conditions. These sophisticated research projects were entries in the Westinghouse Science Talent Search—the best established science competition in our nation. In this competition, students conduct original research projects, write research papers, and submit their work for evaluation. In early spring, 40 of these future scientists are brought to Washington, D.C., where they meet the judges and answer questions about their projects. When the final decisions are made and the winner announced, newspaper and magazine writers descend with photographers and reporters. The students may find themselves instantly famous, as they appear on the covers of national publications. After all the excitement dies down, these students can take advantage of college scholarships which may help them join the prior Westinghouse winners, who have gone on to become accomplished scientists (including five Nobel Laureates), mathematicians, and educators.

RECOGNIZING EXCELLENCE

Much attention is now being paid to disadvantaged youth in our schools—and so it should. As we devote our time to this segment of our population, however, we must also make sure that academically and technically gifted youth receive the help, the recognition, and rewards that they deserve. We must provide opportunities for both groups to discover and refine their skills, skills that could serve the country when these students become the leaders of tomorrow. This is the primary purpose of the competitions that we administer for our students.

Everyone is well aware of the kind of recognition that high school sports heroes achieve. Every student knows that becoming a member of the all-state football team or regional field hockey team brings with it media exposure, college interest, and some local fame. There is nothing wrong with this recognition. In contrast, the academically talented student, the student who may be identified as one of the best science students in the United States, receives a brief announcement on the school's public address system. In proportion, there is something very wrong with this gesture. As a high school teacher interested both in the education of my students and in the future of our nation, the disparity in attention speaks to a dangerous ordering of priorities. Science competitions at all levels can provide a mechanism to shift the balance of attention.

A student who wins a national competition enjoys some fame, receives some money and is told in no uncertain terms that his or her skills are impressive and important. The student also has the opportunity to meet and interact with students nationwide with similar interests and similar talents.

SOMETHING FOR EVERYONE

What of the students who enter but don't come out on top? In my eyes, they too are winners. They have entered a larger arena; they have competed with other students with similar interests and talents. They have learned more science and have stretched and distinguished themselves from the majority who haven't the interest, the energy, or the opportunity to participate. There are in a real sense no losers in the competitions described in these pages.

What if students' interests do not lean toward inventions or a full blown research project? As this book shows, many other opportunities exist. The DuPont Challenge/Science Essay Awards Program requires a bright mind, some research skills, and paper and pencil. The National Bridge Building Contest uses wood, glue and the skills in architecture and engineering to build something that can hold weight—lots of weight.

There are other competitions as well. Some of these—for example, the Science Olympiad—require that students work together and compete as a team. In the Science Olympiad, students from one school compete with those from other schools in a wonderful assortment of challenges. Some students on the 15-member team try to package an egg so that it doesn't break when dropped from the roof. Others enter the "Periodic Table Quiz" or "Name that Organism." Still others try to identify fossils to answer questions in "The Science Bowl." Schools or areas that have not participated in the Science Olympics are soon to be rewarded with some of the best academic fun that exists.

There are paper and pencil challenges for students as well. Students who seem to always get great scores on physics exams may be ready for the Junior Engineering Technical Society Tests of Engineering Aptitude, Mathematics, and Science (JETS TEAMS) competition or the American Association of Physics Teachers/ Metrologic High School Physics competition. In the latter, individuals compete; in the former, a school team works together on the challenges presented.

All the contests described in this paper receive full individual entries below.

WORLDWIDE COMPETITIONS

Students with truly remarkable talents may deserve nomination to one of the international competitions. (See pages 134-145.) Receiving such a nomination may provide that special recognition for which a very talented student is longing. If the student makes it past both the qualifying exam and the final exam, then he or she will be invited to attend a training camp to prepare for an international competition, such as the International Chemistry Camp at the Air Force Academy in Colorado or the International Physics Camp at the University of Maryland. These camps can be the highlight of a student's academic life. For almost two weeks, the students are surrounded with companions of comparable ability. Many didn't know that there were other students with so many similar interests. How many students in one school will be ready (and able) to spend five hours in a row working on a physics problem? At the physics training camp, all 20 students are eager to do so. At an individual school, how many students could absorb a semester's worth of college chemistry in four days? At the chemistry camp, all 20 students revel in the intensity of this instruction.

Ten students—five members of the chemistry team and five of the physics team—represent the United States in the international competitions. Students have traveled to Sweden, England, East Germany, and Poland in recent years. Each year the competition has a unique flavor. Yet each year well over 100 students from around the world find that physics or chemistry is an international language. Students who do not share a common spoken tongue find that they can share with each other problems, puzzles, and a way of viewing the world. These students, who may well one day become colleagues, have through the olympiads gained the opportunity to enjoy each other's company while in their teenage years. The U.S. chemistry and physics teams will soon be visiting Holland, France, Cuba, and Finland to participate in the international competitions. What a wonderful aspiration for our students—if they know enough, can apply their knowledge well enough, and work hard enough, they can represent the United States while still in high school.

TRY IT—YOU MAY LIKE IT

In this book is described a competition for every interested student. Teachers, at the least, should expose students to the range of competitions; of course, competing should never be required. Students could then embark on investigations if they choose to try to become eligible for the competition of their choice. There are obstacles, and it would be foolish not to acknowledge these. One of these obstacles is the teacher, swamped with other responsibilities, who may be afraid that s/he would not know how to help a student entering high-level competitions. Another obstacle is students' lack of interest, due to the other responsibilities they are trying to meet and the fears that they harbor that they may not have the skills necessary to compete. A brief word on each may prove helpful.

It is true that some of these competitions seem intimidating to many teachers. Many do not think that *they* could build a battery-operated device. Some of the sample problems in the Physics Olympiad are daunting to very talented physics problem solvers. Unfortunately, the simplest solution to this dilemma is to toss the application in the garbage in order to hide one's own fears from students. There is another better alternative: let students give the physics problems a try. Let students try the research project for the Westinghouse. When the student runs into trouble, use some local resources.

Every town has some company or some individual interested in electronics. Help the students set up an appointment and get some assistance. Every town is within telephone reach of a college or university. Ask to speak to the chair of the chemistry or biology department. In almost every postsecondary institution, professors are anxious to help an interested high school student. (In the rare exception where this is not true, teachers might ask the department chair [or the president] to explain why secondary students are not encouraged to attend the institution.) Similar suggestions provide materials for the project. Many local companies are pleased to assist a motivated student. If no such company exists locally, write to a larger one. The few dollars it spends on diodes and batteries may be negligible in comparison to the good will that the company derives from its involvement. I am not offering new or unusual ideas. Some of the schools that annually graduate Westinghouse winners have university personnel or industrial scientists serving as advisors to the students.

GETTING HELP

There is no shame or dishonesty in a student receiving assistance as long as that student acknowledges the help. Science teachers do not have to be in there alone. The entire society benefits from successful students. People outside the school community surprise teachers and students alike with their interest and assistance.

I have tried to ascertain from my own students why many of them will not enter a competition. One of the reasons is frightening: "All the smart kids in the country will be trying; I don't stand a chance." What is the message here? What if we extrapolate this comment to the world community? What if students intimidated by their fellow Americans are also intimidated by the Germans and the Japanese? We simply cannot permit such an attitude. We must instill confidence in our students that other students are not simply "smarter" than they are, and that it isn't just the "smart" students— whatever that means—who enter and win competitions. Students who are willing to risk the hard work necessary to be a winner have a good chance of making the finals and winning.

In this regard, we can take our lead from successful sports programs. Although most high school and college athletes realize that they will not be able to make their living playing professional sports, they still enjoy the involvement, the improvement of their skills, the challenge to do their personal best. By removing some of the intellectual mystique, we can encourage this risk-taking in academics as well. In the case of the Duracell competition, have students mentally conceive of a battery-operated device. Tell them not to worry about building it. Even tell them that the idea for the invention is more difficult than building the invention. Try to demystify the act of inventing something. After students come up with ideas, have them help each other. Although the Duracell competition does not allow joint entries, *help, properly acknowledged,* is fine. Let the motivated students pursue their interests. Or start with a team activity, like the Science Olympics—a great place to build confidence. Have students try some of the activities in the classroom, once again, giving students a chance to compete and a chance to succeed.

TROUBLE WORTH TAKING

Administering a competition for a state, a country, or the world is an expensive, labor-intensive venture to which private industry, national scientific organizations, teaching organizations, and many dedicated individuals each contribute. The events happen because the people involved believe that the competitions have a positive impact on students and their education. But competitions can succeed only when students get involved. Teachers should encourage them.

As an advisor to the U.S. team for the International Physics Olympiad, I, along with the team, recently had the privilege of meeting with President George Bush in the Oval Office. An American student had placed first among 150 competitors from 30 nations. The message that I shared with the President was that there were many other academically talented youth in our nation, and many of them do not know that anyone cares about them. I urged the President to try to find ways to encourage these fine students, because in them lies our hope for the future.

As science educators, we must accept this mission and find ways to encourage our students. The vehicles for encouragement exist. Some of them are described in the pages that follow.

Arthur Eisenkraft, physics teacher and science coordinator at Fox Lane High School, Bedford, New York, serves as chair of the Duracell NSTA Science Scholarship Contest and as academic director of the U.S. Team for the International Physics Olympiad.

SCIENCE NATIONAL

AMERICAN ASSOCIATION OF PHYSICS TEACHERS (AAPT)/METROLOGIC HIGH SCHOOL PHYSICS CONTEST

Background: First held in 1985, the contest is sponsored by the American Association of Physics Teachers and by Metrologic Instruments, Inc., to encourage interest in physics. Metrologic contributes the prizes.

Participants: In the past, some 2,100 students have participated in the competition. There are two divisions: Division I is for first-year high school physics students, and Division II, for second-year high school and/or Advanced Placement physics students.

Awards: Metrologic Instruments provides a visible diode laser and accessories to each of the 30 winning schools. In addition, T-shirts are awarded to participating students from the winning schools as well as to those from schools that place second. Every student who participates is recognized by a certificate.

When and Where and How Much: Contests are organized in 15 geographical regions of the United States and Canada. The deadline for ordering application forms, which teachers must do, is early March. The contest itself is held in late April. The fee of $1 per entrant is paid by the school.

Nature of Contest: The contest in each division is based on a multiple-choice test. Because the scores of a school's top four students are compiled to make up the school's score, each school must enter a minimum of four students.

Teachers/Mentors' Comments:

1. One teacher writes proudly, "My students won this [contest] each of the past two years. We have participated in this contest only twice.
2. Another encouraged students to participate in contests through "extra credit grades" and alerting students that colleges will be "impressed."

For Further Information: Write to Ed Gettys, American Association of Physics Teachers/Metrologic High School Physics Contest, Department of Physics and Astronomy, Clemson University, Clemson, SC 29634-1911. Telephone (803) 656-5304.

APPLE COMPUTER CLUBS NATIONAL MERIT COMPETITION

Background: The competition was started in 1984 as an extension activity for the national network of Apple Computer Clubs. In 1989, the sponsors were Apple Computer, Inc., Beagle Brothers, Claris Corporation, McGraw-Hill, Minnesota Educational Computer Corporation, Microsoft Corporation, National Geographic Society, Peter Li, Inc., Scholastic, Inc., Sunburst Communications, and Softdisc.

Participants: The competition is open to all Apple Computer Club members and their advisors across the country, and club membership is open to all students, K–12. A club must be registered under the name of an official advisor, who is usually a teacher. Each year, hundreds of students enter the competition.

Awards: Thirteen national first-prize winners receive Apple computer systems for themselves or for their schools and an expenses-paid trip to Washington, D.C., to exhibit their projects and compete for the grand prize—a Macintosh system-enhanced personal computer and a basic work station unit.

When and Where and How Much: In order to compete, students must send in to Apple Computer $19.95 and applications for membership in early April. The entry fee is included in the Apple Computer Clubs membership. The competition is held in June.

Nature of Contests: To become a national first-prize winner, entrants must achieve top scores in the three categories of grand prize competition described below. Although Apple Computer Clubs has competitions in three categories, only one of these, Category 2, "Basic Programming," is geared toward students at the secondary level. In this competition, students create original programs in any computer language. Programs may be written by individual students or two-student teams. Category 1, "Community Service," is geared toward middle school students and focuses more on community service than computer science. Category 3, "Computers in the Curriculum," is for teachers.

In the basic programming category, past winners include a student who wrote a program that allows teachers to input achievement test scores for the program to analyze and convert to lists of remedial skill lessons. Another student wrote a program that allows the user to draw a shape either free hand or by choosing any of the geometrical shapes offered. The judging criteria for this competition are performance, documentation, originality and creativity, usefulness of the program, and neatness and packaging.

Teacher/Mentor's Comments: One teacher summarizes, "I assisted in overall presentation and the development of the human interface. I also proofread and advised on the documentation. Competitions such as this are great motivators for the truly inspired programmer." The teacher believes that the "recognition offered by a win helps greatly" when students apply to college.

For Further Information: Write to Mary Fallon, Apple Computer, Inc., 20525 Mariani Ave., Cupertino, CA 95014. Telephone (408) 996-1010.

DOW PRESENTS:
THE ART OF SCIENCE

Background: The Art of Science was started in 1986 and is supported by The Dow Chemical Company and sponsored by the New York Academy of Sciences and the National Science Foundation.

Participants: In 1989, 466 students from 46 states and the District of Columbia entered the contest. There were 50 finalists. Full-time students in grades 10–12 may enter. Entrants must be sponsored by a teacher who will certify the originality of the student's art.

Awards: There are several awards: one first-place scholarship ($1,000); four certificates of special merit and honorable mention with a prize of $500 each; and certificates of participation for all finalists. The finalists' works are exhibited at the New York Academy of Sciences gallery during National Science and Technology Week; the exhibition then goes on a national tour for a year.

When and Where and How Much: Students must submit two copies of the official entry form, which includes a 150-word statement about the art, and the art itself, by the deadline (usually in December or January). The New York Academy of Sciences notifies sponsoring teachers of the winners in February. The works selected are then exhibited as described above, and the sponsors publish a brochure announcing the winners and containing reproductions of several of the works of art. Entries must be original; the teacher sponsoring the entry must certify it as such.

Nature of Contest: Students create an artistic interpretation of a scientific concept. By encouraging students to express their views of science through paintings, drawings, prints, photographs, or computer-generated graphics, they learn, according to contest spokespeople, how to "bridge the traditional chasm between scientific and artistic vision." The work of past finalists included an air-brushed and colored-pencil interpretation of Schrödinger's electron wave equations and a detailed etching of layered images representing the evolution of the species. The judges look for depth of thought, originality of expression, and excellence of technique.

Judge's Comments: One judge explains how students present their projects: "Just the work [is] there, with written explanation—this is best in an art competition like this." The qualities for which he looks are "imagination; successful and intelligible connection of science and art; [and] talented execution. [The] degree of ambition...is important, but [the] final project is primary."

For Further Information: Write to Talbert B. Spence, Director, Educational Program Department, New York Academy of Science, 2 East 63rd St., New York, NY 10021. Telephone (212) 838-0230.

THE DUPONT CHALLENGE/ SCIENCE ESSAY AWARDS PROGRAM

Background: The DuPont Challenge/Science Essay Awards Program, which began in 1985, is sponsored by the DuPont Company in cooperation with the General Learning Corporation.

Participants: The participants, students in grades 7 to 12, must be currently enrolled in a school in the United States or Canada or in a U.S.-sponsored school abroad. About 11,000 students participate: Those in grades 7 to 9 compete in the junior division; in grades 10 to 12, the senior division.

Awards: Each division awards several prizes in the form of educational grants: First-place winners receive grants worth $1,500; second-place, $750; and third-place, $500. In addition, the 24 students who earn honorable mentions receive $50 in cash each, and the sponsoring teacher receives a $500 educational award. First-place winners, along with one parent and the sponsoring science teacher, also receive an expenses-paid trip to the NSTA national convention in April.

When and Where and How Much: All essays must be mailed by mid-January to the Science Essay Awards Program (address below). There are no entrance fees.

Nature of Contest: Participants must write an essay, 600 to 1,000 words in length, on the scientific topic of their choice, explaining how this topic has—or will have—a dramatic effect on society. Students may write on any topic in science they choose; they may select a topic that falls within, without, or across the following fields—the physical, life, Earth, or biomedical sciences; technology; and the environment. The contest's informational poster suggests various topics, but these should be viewed as guidelines and suggestions rather than requirements. This essay must be written exclusively for the current year's contest.

For Further Information: Write to Andrea Solzman, Science Essay Awards Program, 60 Revere Drive, Suite 200, Northbrook, IL 60062. Telephone (800) 323-5471; Illinois residents, (708) 205-3073—collect.

DURACELL NSTA
SCHOLARSHIP COMPETITION

Background: The contest started in 1983 with joint sponsorship by Duracell USA™ and NSTA.

Participants: All students in grades 9–12 may enter.

Awards: There is 1 $10,000 scholarship; 5 of $3,000; 10 of $500; and 25 cash awards of $100. Teachers of winners receive computers and NSTA publications.

When and Where and How Much: Five to six hundred entrants each create and build a "working device" powered by one or more Duracell batteries and send an official entry form and a photograph, a two-page description, and a wiring diagram of the device to NSTA by late January. One hundred finalists selected by a panel of scientists and science educators are notified by the end of February. Finalists ship their devices to Duracell Headquarters in early March. All winners are notified by mail, except the top six winners, each of whom learns of his/her achievement by telephone by mid-March and receives an expenses-paid trip with his/her parents and science teacher to the NSTA National Convention in the spring. Winners must be present at the awards ceremony to receive their awards.

Nature of Contest: Entrants' devices must be original creations built by students (and so certified by signatures of the student, parent, and teacher). Students spend anywhere from mere hours to a year preparing their devices. The device should be practical: It should make life easier, or educate, or entertain, or warn, or provide light, sound, or both. But "practicality" can include purely entertaining devices: Past winners have included a "robotic" insect and a "magic" mirror with a psychedelic, sound-activated light display. Over the years, among the winners' inventions were the following: an electric sundial, a talking ruler, an automatic liquid soap dispenser, a cricket that chirps the temperature, a bird-weighing scale, and lighted sneakers. (See Arthur Eisenkraft's paper above in this section.)

Entries are judged on the basis of creativity (30 percent); energy efficiency (30 percent); practicality (30 percent); and clarity of writing (10 percent), including spelling and grammar, in the two-page essay that accompanies each device.

Students' Comments:

1. Writes one, "The competition provides a complete design cycle for someone. That is something most engineers don't get until they're seniors in college."
2. "The competition showed me that you have to be a jack-of-all-trades and master of some," comments another. "Along with engineering, I plan to study writing, business, and trades."
3. A third emphasizes gratitude: "My parents gave moral and financial support. My father taught me years before how to actually make things, and he gave advice on how to go about making the parts needed for my project. My chemistry teacher

came to me about this competition; without that help I would never have won. Two teachers helped me by loaning me books on basic electronics and gave me advice on how to tackle electronic problems. A plastics retailer donated materials for my project."

Teachers/Mentors' Comments:

1. One teacher writes that s/he helped "very little since electronics is my weakest area. I gave [the student] access to my school's equipment. (Actually, I learned how to use some of it from him!) I gave him some class time to work on it. I also told him the parts I thought his report should cover and proofread his drafts."
2. On the other hand, another says that "Some of the students need a great deal of help from the very basics of electronics. Others need an idea for a project; others merely require a little encouragement. Their weaknesses lie in [not] being able to communicate [about] their projects on paper. Other problems arise when [students] try to make their projects small enough or adaptable enough for practical use."
3. A third says, the contest offers a "good opportunity for those students who have the talent to tinker and build to conceive and design a device as simple or as complex as they wish. Each student has different talents; this competition/experience allows those students whose talent is in this area to shine."

Judges' Comments:

1. One emphasizes the thoroughness of the evaluation: "Each application is looked at by two pairs of people independently. They come to an agreement on a ranking. Obvious high scorers and low scorers are separated. Those in the middle are usually reviewed by four more judges at least. Lots of discussion also takes place, drawing on the expertise of the various judges, that is, chemistry, electronics, scuba diving, model airplanes, etc."
2. Another answers the question about qualities sought as "in order of importance, originality, understanding of the science or technology involved, ingenuity (creativity) in executing the idea, and practicality. As I see it, student effort is not as important as the quality of the finished product."

For Further Information: Write to Duracell NSTA Scholarship Competition, 1742 Connecticut Avenue, N.W., Washington, DC 20009. Telephone (202) 328-5800.

ESTES ROCKETRY CONTESTS

Background: Estes Rocketry began sponsoring contests in the early 1960s. Of Estes' three current contests, only Design of the Month and Free Plan are suitable for high school science students.

Participants: Anyone in the country interested in rocketry may enter.

Awards: Winners of the Free Plan contest receive a merchandise voucher good for $100 in Estes products, as well as a certificate. Winners of the Design of the Month contest receive $75 in Estes products and a certificate.

When and Where and How Much: For both contests, participants submit a new design for a rocket. Designs must be the entrants' own and must specify only parts available from the Estes' catalog or made by the rocketeers themselves. Plans must include complete, accurate parts lists. Entrants should actually build and test-fly model rockets to prove their stability in flight. The actual models should not be sent to Estes (they cannot be returned), but judges encourage participants to send photographs along with their designs. For the Design of the Month Contest, entries must reach Estes during the calendar month in order to be entered in that month's competition. Winners of both contests are selected by a panel of judges and are notified by mail. Contestants may enter as many designs as they wish. All entries become the property of Estes and cannot be returned. All winners must meet the standards established in the National Association of Rocketry Model Rocketry Safety Code.

Nature of Contests: The two contests are quite similar. The judges are looking for innovative designs on any type of model rocketry—rockets, booster-gliders, launching or recovery devices, or other designs. The designs should be original and different, but workable, something new that other rocketeers can build and use successfully. Judges place a high premium on accuracy and neatness.

For Further Information: Write to Bob Cannon, Estes Rocketry Contests, Estes Industries, 1295 H St., Penrose, CO 81240. Telephone (719) 372-6565. Estes requests mail-in inquiries only.

FOUNDATION FOR
SCHOLASTIC ADVANCEMENT
SCIENCE COMPETITION

Background: Founded in 1987 by Kenneth Chern, a medical school student at Northwestern University, the concept for the competition originated among a group of high school students who graduated with him from W. T. Woodson High School in Fairfax, Virginia. Funding for the contest is through school entry fees. There are no sponsors.

Participants: Any junior high school or high school student may enter. Approximately 600 schools participate in the nationwide competition, and schools are ranked on the basis of team scores.

Awards: There are several levels of recognition, with awards for first-, second-, and third-place winners. The awards include ribbons, medals, and, for the top-scoring schools at the national level, plaques. Medals for the top team are awarded directly to the students on that team.

When and Where and How Much: The deadline for the science competition is mid-March; for the math competition, it is mid-November. The school entry fee is $25 per school contest. There is a $30 fee for the math contest.

Nature of Contest: In both the science and math contests students compete by taking multiple-choice tests. The science competition has three categories: biology, chemistry, and physics. In contrast to the math contest, which has a team format, the science contest is based upon individual competition. Finalists are the 10 individuals whose scores are highest in the science contest. In the math contest, the top team is the finalist. Teachers from the schools of participating students serve as judges. Three-person teams compete for three rounds each in the math contest. Entrants compete individually on the first round. The other two are cooperative efforts. In the science competition, which has an individual format, the score of the top contestants becomes the team score.

Judge's Comments: Writes one respondent, "The idea behind the Foundation for Scholastic Advancement is to encourage, not tear down, confidence about science. By making questions fun to read, and concise, we attempt to not only review certain skills but teach in the process."

For Further Information: Write to Kenneth Chern, President, Foundation for Scholastic Achievement, 5003 Mignonette Ct., Annandale, VA 22003. Telephone (703) 978-5704.

JUNIOR ENGINEERING TECHNICAL SOCIETY TESTS OF ENGINEERING APTITUDE, MATHEMATICS, AND SCIENCE (JETS TEAMS)

Background: Started by Junior Engineering Technical Society in 1979, the contest is funded by a number of businesses—Fisher–Scientific, the Hewlett Packard Company, the John Fluk Manufacturing Company, the Presidential Classroom, the Small Computer Company, and *Who's Who Among American High School Students*.

Participants: Competing at various levels of the contest are more than 15,000 9th to 12th grade students from private and public schools selected according to individual high school coaches' requirements. School teams have from six to twelve members; the state teams selected from these teams to compete at the national contests have twelve members.

Awards: In the past, they have included science equipment for the schools and scholarship money for the students (including funds to attend a two-week internship in the "Presidential Classroom" in Washington, D.C.).

When and Where and How Much: At national and state levels in 1987–1988, teams competed at sites at 60-plus colleges, schools, and businesses in 29 states; in 1988–1989, at 79 such sites in 40 states. Fees (in the neighborhood of $8 per student) are set specifically by the competition hosts (colleges, schools, and businesses) at the state level; national teams pay no fees, which are assumed by the host institution. Applications are due for state teams in mid-January; contestants at the national level enter by invitation when the state competitions are over. State competitions take place between early February and late March; national, in late April.

Nature of Contest: Junior Engineering Technical Society Tests of Engineering Aptitude, Mathematics, and Science competitions at the state level comprise tests in biology, chemistry, computer fundamentals, English, mathematics, and physics. State-level competitors take 2 tests apiece for 40 minutes each; the school's team score is the sum of the average scores for each subject. At the national level, students divided into state teams face a variety of real-life engineering problems and one technical writing sample during an open-book test. Although the students work together, each team turns in only one set of answers. State competitions are based on multiple-choice tests. State tests are graded by machine. National tests are graded by judges and official scorers at the organization's headquarters.

Contest organizers believe preparing for and participating in JETS TEAMS may raise SAT scores and may help students "grasp and master" certain academic concepts in a college-preparatory curriculum.

Student's Comments: One participant says, "I would recommend this competition for several reasons. The individual competitions in the first round allow students to demonstrate their knowledge of specific concepts and even provide a basis of speculation on how well the students may [be able to] compete against the same school's students in other science competitions." The student continues, "Students advancing to the national competition are in for a unique experience when they team up with students from other schools and pool their knowledge to solve difficult problems, rather than competing against the other team members and attempting to solve the problems individually."

Teachers/Mentors' Comments:

1. One teacher says that it was not necessary to encourage steadfastness because "we used a minimum of time."
2. Another writes, "The students love contests—they are very competitive. Each year we have very special students who are talented, liked, and lead the way for the others to follow."

For Further Information: Write to Cathy McGowan, JETS TEAMS, 1420 King Street, Suite 405, Alexandria, VA 22314-2715. Telephone (703) 548-5387.

JUNIOR SCIENCE AND HUMANITIES SYMPOSIA

Background: Established in 1958, the Junior Science and Humanities Symposia Program has since grown into a national network of symposia at which students present their original research and receive public recognition for their efforts. The Assistant Secretary of the Army (Research, Development, and Acquisition) sponsors the Junior Science and Humanities Symposia Program through the U.S. Army Research Office and contracts with the Academy of Applied Science to act as the fiscal and administering agency. There are 46 regional symposia sponsored in cooperation with universities and science museums. Scholarships and awards are contributed by regional sponsors, the Academy of Applied Science, the U.S. Army, and other local organizations.

Participants: The level of student participation is immense—more than 10,000 students and their teachers from 3,500 high schools attend regional symposia. The symposia are open to high school students who have been nominated by their schools to be participants in a given geographic area. Participant selection is based on the student's interest in doing research, scholastic average, and related academic activities.

Awards: Many regional sponsors and the Academy award scholarships or other recognition for outstanding student research. The U.S. Army awards grants to be used for equipment or other instructional materials. At the national level, the Army sends the seven first-place winners on an expenses-paid trip to attend the London International Youth Science Fortnight held at the University of London in July. The Fortnight is an international conference of over 300 students brought together for the exchange of scientific and cultural ideas. The Army gives the seven second-place winners $200 scholarships.

When and Where and How Much: Regional sponsors ask local high schools to nominate students to participate in the regional symposia. The top five students from each regional symposium attend the National Symposium, which is held in alternating years at West Point or another university. All expenses are paid by the U.S. Army. The top students from each of the 46 regions present their papers to a panel of judges.

Nature of Contest: Any serious student who displays an interest or aptitude for science is eligible to attend as a participant or, if selected, as the presenter of a paper. The level of sophistication of the regional entries varies a great deal, but at the national level, many papers offer significant contributions. Some past entries have been worthy of a master's degree, according to a contest judge. The Junior Science and Humanities Symposia encourage the papers to discuss the interrelationship of the humanities and the sciences.

One past contestant developed a neural network model of visual processing and performed experiments on the model and on human subjects to test the prediction generated by the model.

Student's Comments: One student, learning of the Junior Science and Humanities Symposia Program through his state science fair, writes that the Symposium, as well as the Southern California Academy of Science, the Committee for Advanced Science Training, and the California State Science Fair "are truly invaluable experiences for any serious science student. The Junior Science and Humanities Symposia Program provided something the high school classroom cannot—an original research experience and university support."

Teachers/Mentors' Comments:

1. One teacher, who says that her students are largely self-motivated, gives some details about how she contributes to her students' successes: "I help with ordering supplies and budgeting. I proofread their written reports. There are *many* drafts for the research students; this occurs as their projects progress throughout the school year." While this teacher recommends that students enter competitions, she cautions, "But not with winning as the end. The best part is students' learning poise, increasing self-confidence, speaking ability, and taking pride in a product and a job well done. Students who do 'win' often get money or computers, etc., which helps further their education. Competitions are OK, but...unfortunately, the system can be misused. I don't think they are yardsticks by which to measure success (or lack of it) in a science program."

2. Another teacher, whose students have done well in several national competitions including the Symposia, reports the cooperation of local colleges and industries, who allowed the students use of their labs and research facilities. Of her own efforts, the teacher writes, "We spend about two advanced placement chemistry class periods doing library searches in an effort to find an area and topic of interest. I keep a file of clippings of topics of current interest that might be of interest for possible research. I might throw out ideas and give examples of former projects, but the students do the actual selection. I help with lab procedures—especially if they involve new techniques or instruments. I also give guidance in the proper form used in writing research papers."

3. Being involved as a teacher/coach takes a great deal of time, writes one teacher, who plans to solicit his school's board by writing letters to school superintendents, principals, administrators, "encouraging them to give science teachers *time* built into schedules to assist students with competitions/projects. Everyone that I know in science education spends a tremendous amount of time on competitions/projects without *any* kind of compensation."

Judges' Comments:

1. One of the judges is especially concerned that, given the level of research done, the papers are the students' own work. "I try to determine whether the student understands the work presented. Virtually all students get help. I try to find out who conceived the project, designed the experiments, did the work, and analyzed the results. I look for attention to detail and consideration of alternatives." He further adds, "I look for evidence of insight into the process of conducting scientific experimentation on the part of the student presenter."

2. Another judge writes that from 1 to 5 percent of the entries are rejected outright because they are "essentially library research essays—no involvement by the student in generating new... information." He adds, sadly, that although some of the proposals were sophisticated, they showed no evidence that the student did the investigation. Asked if he had seen evidence that students who don't win feel unhappy, this judge answers, "Yes. Americans promote the 'number one' syndrome, yet major breakthroughs are *team* efforts by persons from many fields with all kinds of different recognitions other than 'number one.' Programs should play up participation, encourag[ing students] to improve over the years through coaching, to...emphasize the value of *growth* and *progress*. Awards should be made available for acknowledging *contributions* regardless of 'number one.'" He further writes, "Disappointment is a fact of life. We used to use politics as an example of getting involved, but now all candidates play a defensive campaign. Yuk! Athletics, music competitions, speech competitions, and *science* competitions should all promote satisfaction of trying and seeing growth—using the accomplishment to buffer the disappointment. There is indeed something of value besides winning."

For Further Information: Write to Doris A. Ellis, Junior Science and Humanities Symposia National Office, Academy of Applied Science, 98 Washington St., Concord, NH 03301. Telephone (603) 228-4520. Also available for reference is a volume containing the Proceedings of the National Association of Academies of Science, 1989–1990, edited by Shyamal K. Majumdar. To order it, write, enclosing $15 in check or money order, to the Treasurer, Department of Biology, College of St. Mary, Omaha, NE 68124.

NATIONAL ASSOCIATION OF ROCKETRY ANNUAL MEET

Background: The first National Association of Rocketry Annual Meet was held in 1959. Sponsored by the National Association of Rocketry, the meet currently comprises numerous separate events at five levels ranging from regional to sectional. For obvious reasons, the National Association of Rocketry puts special emphasis on adhering to official safety codes and regulations. Funding is provided by a large number of businesses.

Participants: Although thousands compete in the events at all levels, only some 150 participants enter the national meet. Of the three competition age divisions, A, B, and C, this description covers only division B, for students 14 to 18 years old. Division A is for pre-high school students; division C is for participants 19 years old and older. All contestants must be members in good standing of the National Association of Rocketry. They must complete and sign an official entry form and have it countersigned by a parent or guardian. Participants must also present to the contest director their National Association of Rocketry Sporting License upon entering a competition.

Two or more members may enter a contest as a team, but they must be registered as such with the contest board. Team membership may not be changed during a contest year, and teams' rockets must fly in the age division of their oldest member. Contestants may not compete both as individuals and as team members, nor may participants be members of more than one team. A group of the National Association of Rocketry members may compete as a chartered National Association of Rocketry section, if all members are in good standing. A contestant may not enter a competition as a member of more than one section.

Awards: Winners receive trophies.

When and Where and How Much: All levels of the National Association of Rocketry competition fall within the contest year, which runs from July 1 to June 30. The national meet is held in August in a different U.S. city each year. Each level of competition is assigned a contest factor, which is used to determine the number of points scored at the meets. Contestants may compete only in as many meets as give them a total of 12 or fewer contest factors, not including the national meet. Each event is also given a weighing factor, based on the difficulty of the event, which also figures into scores. Competition points are awarded as follows: 10 points for 1st place, 6 for 2nd, 4 for 3rd, 2 for 4th, and 1 for making at least one qualified flight. Scores are calculated by multiplying the competition points by the weighing factor, which is then multiplied by the contest factor. Cumulative points for all levels of competition determine the national winner.

Competition begins at the section level, held at local discretion. As the levels go up, the difficulty increases, with the contest factor for local and section meets set at 1, the national meet at 8. In these competitions, two or more sections compete against each other or against all the National Association of Rocketry members in a geographical area. Local and open meets are similarly structured, with certain minor variations; the latter are more difficult and thus have a contest factor of 2. Any entrant in local meets may compete in the open meets.

Regional meets involve members from a wider geographic region than the open meets. Any regional entrant can move on to the national meet. National winners are determined by total points scored at all levels, including their final scores in the national meet.

In addition to these different levels culminating in the national meet, there is also a record trial, in which contestants are given the opportunity to establish or surpass official U.S. and Fédération Aéronautique Internationale model rocket performance records. Participants are given as many opportunities as time and weather permit to make a record flight. Record trials, which can be held at any meet, have no contest factor.

Properly entered contestants may have their models flown by proxy by another National Association of Rocketry member in most contests. Each entry must pass a safety inspection by the safety check officer before each event to be sure that it meets the standard of the sporting code.

Nature of Contests: There are five categories of contests: Altitude, payload, duration, craftsmanship, and miscellaneous events with many subcategories.

Altitude Events. In this category, the altitude and super-roc altitude events each have nine different classes of events based on the permissible total impulse of the rockets' motors. The object of the altitude competition, which is open to any model rocket, is to achieve the highest altitude.

Payload Events. Each of the events in this category has several classes for entry based on the permissible total impulse of the rockets' motors. Open to model rockets carrying one or more standard National Association of Rocketry payloads, these events' purpose is for rockets to carry given payloads as high as possible and to recover them.

• The egg-lofting competition has seven classes of events and is open to models that carry, as a totally enclosed payload, a raw egg. The object of the competition is to carry the fragile payload as high as possible and recover it without damage to simulate the precautions and restraints employed to cushion an astronaut's landing.
• The dual egg-lofting competition is like the egg lofting except that it has four classes of competition instead of seven and the rockets carry two eggs instead of just one.

Duration Events. These events focus on devices that stay aloft for the longest possible time; all but the last are divided into classes based on the permissible total impulse of the rockets' motors.

• The parachute duration competition has five classes of events open to single-staged rockets with one or more parachutes for recovery purposes.
• The streamer duration competition has nine classes of events open to single-staged entries that have a single streamer as the only recovery device.
• The helicopter duration competition has nine classes of events and is open to any single-staged rocket that uses the principle of autorotation as the sole means of recovery.
• The super-roc duration competition has nine classes of events open to single-staged models with a body length no less than the minimum allowed for the class. Contestants try to achieve the longest possible duration without damaging the structural integrity of their models.
• The egg-lofting duration competition has six classes of events open to single-staged rockets that carry, as a totally enclosed payload, a raw egg. The object is to carry the payload for the longest possible time without damaging the egg.
• The booster glider competition has nine classes of events open to any model rocket, one portion of which returns to the ground in stable, gliding flight supported by aerodynamic lifting surfaces that sustain that portion against gravity.
• The rocket glider duration competition has nine classes of events open to any single-staged model rocket that returns to the ground in

stable, gliding flight, supported by aerodynamic lifting surfaces that sustain it against gravity.

- The flex-wing boost glider duration competition has nine classes of events open to any model rocket, one portion of which returns to the ground in stable, gliding flight supported by flexible aerodynamic lifting surfaces that sustain that portion against gravity.
- The precision duration competition is open to any single-staged model rocket and comprises three events: predicted, set, and random duration.

Craftsmanship Events. These events comprise six competitions: scale, scale altitude, super scale, sport scale, space systems, and plastic model conversion. In all of them, contestants try to produce true scale models of existing or historical guided missiles, rocket vehicles, or space vehicles. In all of them, the models must be accurate, flying replicas that exhibit craftsmanship in construction, finish, and flight performance.

Miscellaneous Events.

- The spot landing competition is open to single-staged models and comprises three events: parachute, streamer, and open spot landing. The object is to land the model so that the tip of the nose cone is closest to a predetermined spot on the ground.
- The drag race competition is open to single-stage entries. Contestants try to achieve the best scores in the triple criteria of quick ignition and lift-off, low altitude, and long duration.
- The radio controlled rocket glider competition is open to any single-staged rocket that is radio controlled during flight and return to the ground. The object is to repeatedly achieve specified flight times and land as close as possible to a designated spot over a series of three flights.
- The research and development competition is open to any National Association of Rocketry member doing research or engineering new developments in which model rocketry plays a primary part.

For Further Information: Write to Maria Stumpe, Headquarters Manager, National Association of Rocketry, 1311 Edgewood Dr., Altoona, WI 54720. Telephone (715) 832-1946.

NATIONAL BRIDGE BUILDING CONTEST

Background: Physics teacher Roy Coleman of Morgan Park High School in Chicago first tested a model bridge at a regular meeting of a group of physics teachers. Impressed, members of the group started the National Bridge Building Contest in 1977. The contest is administered by the National Bridge Building Committee, which is composed of the original group of physics teachers, an engineer or two, a student or two, and a professor or two from the department of physics at the Illinois Institute of Technology in Chicago.

The competition is funded by a small profit made by selling bridge kits to teachers who conduct school contests; by Illinois Institute of Technology alumni; by anonymous contributors; and by Midwest Products.

Participants: Any high school student in grades 9 through 12 who is one of the top two winners in a regional contest may compete in the national contest.

Awards: The participants who place in the top three positions receive prizes such as computers and cameras. In the past, the first prize has also included a half-tuition scholarship to the Illinois Institute of Technology.

In addition, each school receives a trophy for participation, and each student receives a certificate as well as a science trinket.

When and Where and How Much: Usually in early April, the deadline for entry is set each year by the national contest host. The location shifts around the country; most entrants travel to wherever the contest is held—recently, in Chicago, New York, Denver, and Bellingham, WA. Contestants are also permitted to qualify by mailing their entries.

Nature of Contest: Student designers/participants construct bridges, which they then test to destruction, in order to determine efficiency. The most efficient entry wins. Mail-in bridges are tested by an engineer or stand-in student.

Students' Comments:

1. One has mixed emotions: "I felt great, because I won, but when my bridge broke during the test, it hurt."
2. A second recommends the competition to other students because "it stimulated me to think constructively." Nonetheless, this student says, "I doubt very strongly that I will ever become a mathematician or scientist. I am currently on course to be a professional welder or mechanic." The student concludes, "I believe competitions such as bridge building force one to think of the trouble, amount of time, and thought [that go] into designing [the everyday things] around us."

Teachers/Mentors' Comments:

1. One writes, "It is time consuming to construct and test several bridges and we have a fairly high attrition rate. The students who persevere reap the rewards. The intellectual rewards have been inspiring, and [the students] have placed high at both regional and national levels. The upper level competition has involved trips. For instance, we have toured such places as [the] Brookhaven National Lab in New York and the Fermi Lab in Chicago."
2. Another is also enthusiastic: "The students enjoy the contests once they get to the competition, no matter how well they do themselves. They are proud when a fellow classmate wins or [when] the school wins." The teacher continues, "I feel competitions are healthy when not overdone. The competitions offer [the] academically talented and the "tinkerers" a chance to excel. [In addition,] girls have done well in bridge building competitions, with usually at least one in the top five in the state."

For Further Information: Write to Earl Zwicker, Department of Physics, Illinois Institute of Technology, Chicago, IL 60616. Telephone (312) 567-3384.

NATIONAL FUTURE FARMERS OF AMERICA AGRISCIENCE STUDENT RECOGNITION PROGRAM

Background: The program, begun in 1988, is a special project of the National Future Farmers of America Foundation and is sponsored by the Monsanto Agricultural Company of St. Louis, Missouri.

Participants: Members of the National Future Farmers of America organization may participate as juniors, seniors, or recent graduates of high school, or as college freshman enrolled in an agriculture program. All entrants must have excellent grades with an academically demanding schedule that focuses (or focused) on the application of scientific principles and emerging technologies in an agricultural enterprise. At the national level, there were 30 entrants in the 1989 competition.

Awards: There are a number of levels of award and recognition. At the state level, 50 $1,000 scholarships are awarded; at the regional level, there are awards of eight scholarships of $2,500 each. The national winner receives a $5,000 scholarship; the national runner-up, a scholarship for $3,000. All awards are applicable to tuition charges of the school of each winner's choice.

When and Where and How Much: All state-qualifying applications must be submitted to the National Future Farmers of America Center by mid-July. State deadlines are set locally by administrative offices for programs in vocational agriculture education. Regional finalists travel to the National Future Farmers of America Convention in Kansas City, Missouri, in early November. There is a partial travel award.

Nature of Contest: Students are encouraged to conduct agricultural research projects related to their classroom or lab instruction. Winners must submit an application as well as a research paper. Awards are based on originality, creative ability, scientific thought or goals, skill and clarity (together counting 50 percent); academic achievement (35 percent); and school and community activities (15 percent). At the regional and national levels, a committee decides on the finalists.

Judge's Comments: One respondent notes, "This contest gives students the opportunity to think, create, and develop at the highest level. It involves students in exceptional projects, which also aid in teacher growth and improve curriculum. Teachers also get involved, which tends to increase their awareness and interest in a wider area of agriculture."

For Further Information: Write to Carol Duval, Program Coordinator, National Future Farmers of America Organization, National Future Farmers of America Center, 5632 Mt. Vernon Memorial Highway, P.O. Box 15160, Alexandria, VA 22309-0160. Telephone (703) 360-3600.

NATIONAL GEOGRAPHY OLYMPIAD*

Background: The National Geography Olympiad is sponsored by the National Council for Geographic Education, a national nonprofit educational organization. Funding for the competition is derived from the registration fee paid by schools that enter teams of students in the competition. First held in 1986, the National Geography Olympiad operates under the auspices of the National Social Studies Olympiad. The tests participants take are developed by the National Council for Geographic Education.

Participants: Any student from grades 4 to 12 may participate. There are six categories of participants, grouped according to grade level: grades 5 and under (438 teams); grade 6 (411 teams); grade 7 (368 teams); grade 8 (236 teams); grade 9 (154 teams); and grades 10 to 12 (103 teams). Each team has at least 10 students. There are a total of 1,755 teams with about 20,000 individual members.

*The National Geography Bee, a contest for younger children sponsored by the National Geographic Society, has been available for students in grades 4 to 8 since 1988.

Awards: Each team registered, regardless of final standing, receives 11 awards—a medal and 10 certificates—which may be distributed as individual schools choose. At the national level, the team award is a large parchment document listing the names of each student. Each student also gets a medallion.

When and Where and How Much: The contest is held at the participants' schools on any day in the first two weeks of April chosen by the individual school. The school registers the first team for $40 and each additional team for $20. There is no limit to the number of students who may participate, provided that they are placed on teams according to the correct grade category.

Nature of Contest: The contest consists of a 35-minute multiple-choice test, with the level of difficulty and number of questions increasing with the grade levels of participants. The contest for grades 4–6 comprises 40 questions; upper grades answer 50 questions. All tests contain maps and graphs and include skill-related questions (for example, questions deal with direction and distance and ask for interpretation of maps or graphs) as well as questions pertaining to physical and human geography. Students work independently, and each test is proctored and graded by the school. The sum of the top ten scores, recorded on a score sheet and returned to the National Geography Olympiad, is tabulated for all schools, each of which receives a report of the national results.

Teacher/Mentor's Comments: Writes one teacher about competitions such as the National Geography Olympiad, they offer "a means of judging your performance as a teacher with those in other schools."

For Further Information: Write to Joseph Quartararo, National Geography Olympiad, P.O. Box 477, Hauppauge, NY 11788. Telephone (516) 265-4792.

NATIONAL HEALTH OCCUPATIONS STUDENTS OF AMERICA COMPETITIVE EVENTS PROGRAM

Background: The National Health Occupations Students of America Competitive Events Program was started in 1978 and is sponsored by Health Occupations Students of America and a number of professional associations in the health care field. The 26 officially recognized events are organized into 4 categories: Health Occupations Related Events, Health Occupations Skills Events, Individual Leadership Events, and Team Leadership Events. Health Occupations Students of America's Competitive Events Committee regularly monitors the Program in order to review and revise existing events and direct the development and addition of new events to reflect the current high school curriculums. There are several pilot events each year.

Participants: While Health Occupations Students of America offers membership to postsecondary and collegiate as well as secondary students, only the eligibility requirements for secondary students are listed here. Students must

- be members of National Health Occupations Students of America registered on an official chapter assessment roster with assessment paid prior to January 1
- be enrolled in state-approved health occupations education programs
- not receive their high school diplomas (or equivalent) before the state's annual conference
- have been enrolled in high school for two or more years prior to the current year's national conference

Participants in the national competitive events must have competed in and won the same event at the state level. At least half of the members of teams registered by team events must be students who were members of the winning team at the state level.

Awards: First- through third-place medals are awarded in each division.

When and Where and How Much: Participants must register for the national competitive events by mid-May. The events are held in late June during the National Leadership Conference in a different U.S. city each year. Participants must be able to pay their own travel expenses; however, in some cases the states pay travel expenses.

Nature of Contests: Of the 26 events, the following 3 are representative of the various competitions: medical laboratory assisting, dental laboratory technology, and the Health Occupations Students of America Bowl. The other 23 events focus on developing and recognizing personal and leadership skills related to the health occupations. Judges award points for every step in a given procedure and penalize entrants for exceeding the allotted time. The event chairperson may also deduct points for improper clothing or equipment. Teams competing in the national event must consist of at least two members who participated on winning teams at the state level.

Medical Laboratory Assisting. The purpose of the event is to encourage health occupations students to develop and apply the basic entry-level skills and knowledge required. Competitors are timed as they demonstrate professional skills in several of the following areas: hematology, immunohematology, urinalysis, and bacteriology.

Dental Laboratory Technology. The purpose of this event is to encourage students to develop and apply the basic entry-level skills required of a dental laboratory assistant/technician. Competitors are directed to perform tasks selected from a list of 20 procedures, for example, fabricate a maxillary or mandibular individually designed impression tray with spacers on edentulous casts; set up a standardized maxillary tooth arrangement in relation to the occlusal plan and the oblique plane; and repair a maxillary or mandibular full denture that has a midline fracture and replace 2 broken or missing teeth.

Health Occupations Students of America Bowl. This event is intended to stimulate students to work as a team and to test their knowledge on various topics covered in health occupation education programs. Two teams of four members each compete by giving appropriate responses to items presented by a moderator. These may be questions, incomplete statements, or definitions. The team that answers the most questions within a 10-minute period wins the match. Winners are determined by a series of elimination rounds. The last remaining team becomes the first-place winner of a section.

For Further Information: Write to Jim Koeninger, Health Occupations Students of America, 6309 North O'Connor Rd., Suite 215, Irving, TX 75039-3510. Telephone (214) 506-9780 or (800) 321-4672.

NATIONAL JUNIOR HORTICULTURAL ASSOCIATION PROJECTS

Background: Started in 1935 with the cooperation of the Vegetable Growers Association of America, this initially was a "vegetable growing, judging, and identification contest for young people." When the National Junior Horticultural Association was formed in 1939, it took over administration of the contest. The Projects today comprise 15 different events. Only the three suitable for high school science students are covered below. The Association and the Projects are financed through voluntary contributions from some 50 sponsors, including corporations, associations, agencies, and individuals.

Participants: Eligibility varies with each contest, but all participants must be members of the Association. The Horticulture Contest and the Experimental Horticulture Project are open to members between the ages of 15 and 22 and the Photography Exhibition to any member from 8 to 22.

Awards: Most of the national awards are pins, emblems, framed certificates, plaques, gift certificates, and trophies. Some projects also provide trips and/or cash awards. In addition, some states provide trips or other special awards to state winners.

When and Where and How Much: Students who wish to participate must complete an enrollment form and request the appropriate project report forms from the state Association leader. Membership is free, but contributions are welcome. For all of the projects covered here, participants enter at the state level and, if they win, go on to the national finals at the Association's annual convention.

The national office pays the travel expenses of National Junior Horticultural Association youth officers only. Some state associations will help pay travel expenses; students can also get help from other organizations, such as the Future Farmers of America, the 4-H, or county extension services.

Nature of Contests: The three science contests appropriate for the high school level are discussed below.

Horticulture Contest. "A tough contest" is the assessment of contest chairperson Robert Muesing. It consists of a three-part test covering the qualities consumers look for in buying horticultural crops and products. An objective, 80-question exam, which tests the participant's knowledge of plant science, cultural practices, pesticides, horticultural careers, gardening, and other topics, includes a section where entrants must identify 100 specimens. The contest has seven divisions:

1. 4-H—official state club teams of three or four members, one team per state (three high scoring individuals become an official team)
2. Future Farmers of America—official teams of three or four members, one or more teams per state (three high scoring individuals become an official team)
3, 4. Open Teams—one team for ages 15–18 and another for ages 19–22 (open to any three-member team from a club, school, county, or state)
5, 6. Individual—one group for ages 15–18 and one for ages 19–22 (open to any individual not participating in the contest as a team member)
7. Honors—open only to official members of teams placing first in any division of the contest or to national winners in any of the other divisions, so named either as individuals or as official team members

Experimental Horticulture Project. Participants select a specific problem to investigate and discuss with a teacher or leader. They carefully plan and organize their materials and methods and keep detailed records of the results, so that the final report clearly outlines all information obtained. Most projects fall into one of two groups: those dealing with cultural field trials, such as row spacing, variety trials, fertilizer trials, cultivation, irrigation, and the like, or those dealing with laboratory or greenhouse studies, such as the use of growth regulators, plant breeding, hydroponics, the effect of light and other environmental factors on plant growth and reproduction, or vegetative propagation.

The National Junior Horticultural Association state leader, working with a committee, grades all reports on a numerical basis, with a maximum possible score of 1,000 points. Evaluation is based on the type of project, the specific problem involved, the materials and methods used, the results obtained with a conclusion and summary of the study, along with a bibliography of references used. Prospective winners are urged to attend the national convention to be interviewed for additional credit.

Photography Exhibition. Participants submit an entry to their state leaders by mid-July. Each state leader selects one entry from each of three classes and forwards the choices to the project chairperson by early August. Participants enter in these classes:

- Class 1: a single black-and-white photograph (maximum 10 x 12 inches) of a horticultural subject or activity
- Class 2: a single color photograph (maximum 10 x 12 inches) of a horticultural subject or activity
- Class 3: a sequence of four snapshots in black-and-white or color, representing a horticultural event or activity, with a layout plan for final mounting

In general, judges consider subject matter (appropriateness as a horticultural exhibit, aesthetic appeal, viewer interest), composition, and technical quality. Every entry is evaluated on its own merit and receives a designation as outstanding, good, or worthy.

Students' Comments:

1. Writes one, "I came across the object of my photograph by accident. I was on a work-related tour in Florida when I discovered a unique plant growing on the lawn in front of a restaurant. I quickly photographed it, and it became a prize-winning picture."

2. In response to a question as to whether she would eventually end up working as a scientist, another student writes, "Yes. I am currently a junior at North Carolina State University, majoring in ornamental horticulture. I plan to go to graduate school and eventually would like to teach horticulture or work as a county extension agent. This contest was a way for me to participate in my favorite subject—horticulture!"

Teacher/Mentor's Comments: One teacher writes, "We are involved in all aspects of our 4-H'ers programs—we will help in just about any way asked. When we are preparing for a contest, we work approximately twice each week for between two and three hours at a time."

For Further Information: Write to Jan Hoffman, National Junior Horticultural Association, 441 E. Pine St., Fremont, MI 49412. Telephone (616) 924-5237.

SCIENCE OLYMPIAD

Background: Established in 1985 by Jack Cairns of the Delaware Department of Public Instruction and Gerard Putz of the Macomb, Michigan, Intermediate School District, the Science Olympiad is a nonprofit organization that sponsors local, state, and national tournaments of approximately 23 individual and team events. Each year, the Olympiad introduces new events to keep the tournaments interesting and timely. The Science Olympiad is funded by the U.S. Army ROTC Cadet Command, the American Honda Foundation, "TRW," General Motors, the U.S. Army Research Office, DuPont, and the Dow Chemical Company. The competitions follow the format of board games, television shows, and athletic games, in which are represented the various scientific disciplines of biology, Earth science, chemistry, physics, computers, and technology. There are four divisions in the competition, but only Divisions B and C, in which students in grades 6–12 compete at the state and national levels, will be discussed here. No other divisions serve high school students; however, there are two elementary divisions for grades K–6.

Participants: The Science Olympiad estimates that more than a combined one million students participated in the different tournaments in 1989, with about 2,000 contestants at the National Tournament. The secondary tournaments are open to individuals and teams in grades 6–12 who are members of Science Olympiad. A team may have as few as 2 and as many as 15 students. A maximum of 5 9th grade students for Division B and 7 12th grade students for Division C may participate on a team. Most states and regions restrict participation to one team per school; others encourage multiple memberships. Some schools have as many as 10 paid memberships, with all 10 participating in tournaments. The state director notifies the school if multiple memberships are permitted. Membership entitles a team to participate at the first level of competition in that state, whether that tournament is district-, region-, or statewide. The top winners in these competitions are eligible to continue to the next level tournament. A team, composed of as few as two students, may participate in one or all of the events in its division.

Awards: Gold, silver, and bronze medals are awarded for first, second, and third place in each event. Championship trophies go to the school teams that score the most total points. In addition, the American Honda Foundation provides scholarships to 46 gold medal winners.

When and Where and How Much: Participants must register for membership at least 30 days before the regional or state tournament. The membership fee for Division B and C teams is $35. The contestants' travel expenses are paid by the schools or are funded by businesses or local parents' groups. The National Tournament is held, usually in May, at a different U.S. college or university each year.

Nature of Contests: Of the Olympiad's 23 events, only 10 are appropriate for high school students. These 10 are described below.

Bio-Process Lab. This event is designed to test the students' knowledge of processes in the biology lab. They are asked a series of questions that call for interpreting words and pictures and performing tasks, such as comparing the different rates at which pared and intact apples dry out by weighing the apples, making charts or graphs, or simply stating a hypothesis. Students should be prepared to interpret data, use instruments, and make predictions, observations, and inferences within set time limits. The number of questions and tasks varies with each tournament.

Bridge-Building Contest. This event is for individuals or teams of two, who construct their bridges before the competition starts. The objective is to design and build the lightest possible bridge capable of supporting a given load over a given span.

Circuit Lab. Two-member teams receive electrical data from which they must predict current and voltage values along a given circuit. The teams have 40 minutes to solve 3 problems. They might, for example, be given input voltage and resistances along a circuit (both series and parallel) and asked to calculate electrical readings if ammeters and voltmeters were inserted in the circuits. Each correct answer earns points for the team, and the team with the highest score wins.

Computer Programming. Designed to test students' ability to write a computer program that solves a science problem, this event asks two-member teams (although an individual may enter if no other student from that school is available) to develop and test their programs on specified problems within the set time limit. Teams must solve three science problems of varying difficulty using BASIC, a programming language. The winning team is the one that submits the correct solution to the problems in the shortest time.

Metric Estimation. Individual contestants estimate the mass, volume, area, distance, capacity, or temperature in appropriate Standard International metric units of 20–30 objects. The students move from station to station, recording their estimated measurements on an official form, which they submit to the judges after completing the course. Scores are based on the accuracy of the estimates: An estimate within 10 percent of the measured value is awarded 5 points, within 20 percent, 3 points, and within 30 percent, 1 point.

Pentathlon. In this event, teams of two girls and two boys work their way through an obstacle course of both physical and academic events. Schools that are not coeducational may recruit teams of the opposite gender from elsewhere. After surmounting a physical obstacle, such as a broad jump, each team member stops to answer a question from each of five science areas: biology, Earth science, chemistry, physics, and environmental education. Contestants must also run the course while carrying a balloon filled with air, helium, water, or another substance. Each physical and academic obstacle must be overcome before contestants advance to the next.

Periodic Table Quiz. This event is conducted as is a spelling bee: instead of spelling words, however, the contestants answer questions on elements, compounds, and the use of the periodic table. Contestants, who line up, must give a correct answer within 30 seconds in order to remain in the competition. If a contestant cannot answer a question correctly, s/he sits down, and the same question is given to the next person in line. The last contestant left standing wins; the last contestant eliminated takes second place; and the next to the last is third.

Qualitative Analysis. The object of this contest is for 2-member teams to correctly identify the solutes in 10 numbered vials as quickly as possible, on the basis of the solutes' reactions with each other. Each team receives 10 small vials (each containing about 10 ml of an aqueous solution), a list of the formulas of the 10 solutes, 10 empty test tubes and a rack, dropper pipettes, wash water, and a report form. To identify the solutions, the teams use evidence about each solution's color, odor, and/or the results of mixing small amounts of unknowns. Precipitate formation, gas evolution, and heating effects offer other evidence. The winning team identifies the most solutions correctly in the shortest period of time.

Science Bowl. Patterned after television's "College Bowl," this event features four-member panels competing against one or two other panels by answering questions taken from biology, chemistry, Earth science, and physics textbooks. There are two kinds of questions. For toss-up questions, the person who signals must answer within five seconds without conferring. For bonus questions, awarded for a correct response to a toss-up question, panel members may consult, after which their designated spokesperson answers. The high scorer in the preliminary round moves on to the semifinals, and the high scorer there qualifies for the finals, where the highest scorer takes first place.

Sounds of Music. In this event, three-member teams build musical instruments, describe the physical principles behind the instruments' operation, and perform musical selections on them. Each team must build at least three instruments from materials not commercially designed for producing music or sound. Members of the team play the instruments, which are evaluated on range and sound quality. Judges ask contestants to describe the principles behind the devices and their design and construction. The team then performs two pieces of music—one required and one freely chosen. Points are given for the creativity, variety, and quality of work on the instruments; the range of notes and quality of sound; the performance of both tunes; and the participants' knowledge of the theoretical basis of the instruments. One winning team came up with a hose-a-phone and a slide parsnip. Sounds-of-Music team members must be prepared to be part of the Science Olympiad Orchestra and perform "My Country `Tis of Thee" as part of the opening ceremony.

Student's Comments: A national gold medalist in the Qualitative Analysis event who thinks he will enter the physics or engineering fields says he "worked and prepared for the competition at school mostly and sometimes at home. The school provided any materials and facilities that they possessed and were needed for the competition."

For Further Information: Write to Science Olympiad, 5955 Little Pine Lane, Rochester, MI, 48064. Telephone (313) 651-4103.

SPACE SCIENCE STUDENT INVOLVEMENT PROGRAM

Background: Started in 1980 by cosponsors NASA and NSTA, the Space Science Student Involvement Program is designed to interest students and teachers with diverse backgrounds in aerospace science and technology. Thus, the seven current competitions, of which four are for high school students, range from writing articles and advertisements to proposing experiments that could be performed on a space station. Contests change from year to year.

Participants: In 1989, more than 1,500 students and teachers participated in the program. While the Space Science Student Involvement Program includes contests for students in grades 6–8 as well as those in grades 9–12, only those for secondary school students will be covered here. The program is open to students in the United States and its territories. Teachers/advisors do not have to be members of NSTA.

Awards: For the Space Station Proposal and the School Newspaper Promotion: All students and their teachers receive certificates of participation; students who earn honorable mention and their teachers receive certificates of recognition; national winners and their teachers receive plaques and an all-expenses paid trip to the National Space Symposium, traditionally held in Washington, D.C., in September. In addition, national winners are eligible to compete for a student Space Science Foundation scholarship. For the National Aerospace Internships: All students and their teachers receive certificates of participation. National winners and their teachers receive a one-week expenses-paid educational visit to a NASA research center. For the Mars Settlement Illustration: All students and their teachers receive certificates of participation; for students who earn honorable mention and their teachers, certificates of recognition; for the students in each category whose entries are named national winners, a cash award, and for their teachers, resource materials. Winning entries are mounted in a national traveling exhibit.

When and Where and How Much: Students and their teachers must complete the official Space Science Student Involvement Program entry form and send it in by the deadline, which is usually in mid-

March. There is no entry fee. Both student and teacher must certify that the entry is the student's own work, with any outside assistance acknowledged on a separate sheet. Space Station Proposals, along with entry forms, must be sent to the appropriate regional director by the above-mentioned deadline. National Aerospace Internship proposals and entry forms are sent to the National Science Teachers Association by the deadline. The School Newspaper Promotion requires that all entries be published in the student's school newspaper by February 1 and that two copies of the newspaper in which it appears and the entry form be sent to NSTA by the deadline. For the Mars Settlement Illustration, participants must send the art and the entry form to NSTA by the deadline. Students and teachers are notified of their status by late April.

Nature of Contests: There are four secondary-level Space Science Student Involvement Program contests.

Space Station Proposal. For this contest, students propose experiments that theoretically could be performed on a space station. The proposals should deal with problems related to space science, such as the use of space microscopes, microgravity and human health, or computer-staged events. Judges award points for scientific validity, suitability as a space station research facility, creativity, originality, organization, and clarity. Regional winners proceed to the national competition. Past winners have proposed experiments on the effect of microgravity on the vital lung capacity of the human respiratory system and the effect of space environment on the proliferation of resting, activated, and malignant T(213) lymphocytes.

National Aerospace Internships. In this contest, students may submit proposals on one of two topics: experiments that theoretically could be performed in the Wind Tunnel Testing Facility at the NASA Langley Research Center in Hampton, Virginia, or experiments that theoretically could be done in the Zero Gravity Research Facility at NASA's Lewis Research Center in Cleveland, Ohio. Students are given the physical specifications and capabilities of the facilities and are asked to come up with proposals with scientific validity, suitability to the research facility, creativity, and organization, and clarity. A previous winner in the wind tunnel category proposed control surface testing on a forward sweep prototype aircraft. One zero-gravity researcher proposed measuring the tensile strength of pure water.

School Newspaper Promotion. This contest allows students to demonstrate their journalistic and design skills in either of two categories: They may write a news or feature article about some aspect of the Space Science Student Involvement Program or space science and students, or they may design and write an advertisement for the current year's program. Students submitting articles should note that judges look for journalistic style, including creative angle and interest to the reader, accuracy and content, and layout and design. In the advertisement entries, judges rate layout and design, accuracy and enthusiasm, and execution and craftsmanship.

Mars Settlement Illustration. In this contest, students research and illustrate their concepts of the first human settlement on Mars. Here, again, are two categories for entries—general illustration and diagram/schematic. In the general illustration, students must illustrate the overall settlement, which must support 10 people and show life support and transportation. A diagram or schematic must show the technical features of the settlement. The medium must be suitable to two-dimensional work; no three-dimensional work is acceptable. Contestants must also write two to five paragraphs describing how they arrived at their illustrations, their scientific justification of the settlement's design and feature, the media used, the features of the work and any other relevant information.

Students' Comments:

1. One winner answered at length when asked if she thought she would eventually end up working as a mathematician or scientist: "Currently, I am employed at General Electric Company in Valley Forge, Pennsylvania, working as a systems engineer on the Space Station Program. I studied engineer[ing] and physical science at Villanova University (graduated '88). I feel the expos[ure] and experience during the contest guided me into my present career. I enjoyed creating the experiment, and I *love* my job with GE/Space Station Freedom. Through the contest, as a regional winner, I was able to meet an astronaut, which was also very exciting. I say this because I now work with them! Going from the position of student to engineer has been very interesting as I step back and look at it. I have become [what] I've wanted to be in high school, a scientist whose [invention may] orbit the Earth. Currently, I am looking to further my education with a Master's in Engineering from Penn State. *Learning never stops!*"

2. Another student explains, "I designed an experiment that could possibly be sent up in the space shuttle. My experiment tested the ability of a mold (*Pilobolus crystallinus*) in shooting its spores accurately in zero gravity. [The idea] came about from a discussion with my science teacher. He described a similar experiment where students used the same procedures. I realized that a zero gravity environment would add a different twist to the experiment."

For Further Information: Write to Helenmarie Hofman, NSTA, Department of Science, Space, and Technology, 5110 Roanoke Pl., Suite 110, College Park, MD 20740. Telephone (301) 474-0487.

SUPERQUEST

Background: SuperQuest is the only super-computing competition in the United States intended exclusively for high school students. Its purpose is to foster creativity in devising computational solutions to a scientific problem. After its first year (1988), the sponsorship of SuperQuest changed from the supercomputer subsidiary of Control Data Corporation, which first funded the competition, to the Cornell National Supercomputer Facility. In 1989, with support from the National Science Foundation and IBM, the Cornell National Supercomputer Facility conducted the contest.

Participants: In 1989, SuperQuest had 71 team entries. Full-time students may enter the competition only as part of a three-to-four member team, and each team must have a teacher/coach who carries a full teaching course load. Any accredited secondary school in the United States and its territories may enter more than one team. There is no minimum age limit, but members of the team must not have passed a high school equivalency exam or had their 19th birthday before the application deadline, and no more than one student member of the team may be in grade 12. If a school sponsors more than one team, no student member of any team may be a member of another team or participate in another team's project. Two or more schools may jointly sponsor one team. The four schools that sponsor winners in a given year may field teams in the following year, but the individual team members, including the coach, are not eligible to compete. No more than one of the winning teams from the current year may be from any of the four schools that won in the previous year. All entrants must be U.S. citizens or legal residents.

Awards: The four teams selected as winners receive an IBM work station configuration for their home schools, one year of access to the Cornell National Supercomputer Facility supercomputers, and attendance at the SuperQuest Institute in Ithaca, New York, for which teacher/coaches receive a $3,000 stipend and students receive a $1,000 stipend. Awards for the "best student paper" are $1,500 for the winner and $1,000 for the runner-up. In addition, the school that wins the "best school" competition receives continued access for another academic year to the Cornell National Supercomputer Facility. Winning team members and their coaches must be available to spend three weeks during the summer in residence at Cornell University. A team that cannot send three students and their coach to the SuperQuest Summer Institute is disqualified. A team's school board must authorize the team's participation with the understanding that the school must provide space for a scientific work station configuration and agree to actively use the work stations and the Cornell supercomputers in its curriculum if the school's team should win the competition. The winning school is provided access to Cornell's supercomputers by Cornell staff members who arrange an interface with Cornell's computer system. No more than one team per school may be selected to attend the SuperQuest Summer Institute.

When and Where and How Much: Entering and competing in SuperQuest goes in four phases. The first is registration: Teams must submit an official team entry form by the March deadline. In addition to the team entry form, teams should include a 200-word abstract for each student's project. Coaches receive a postcard confirming the team's registration, complete with a team identification number.

In the next phase, by mid-April, teams submit a project report, which should include a school administration authorization form and a statement outlining how the school will use the work stations and access to the supercomputer. There should also be a completed biographical form for each team member and the coach, as well as a project report describing each student's progress.

The next phase of the contest is review and notification, during which an independent review committee appointed by the Cornell National Supercomputer Facility evaluates each report. Around mid-May, the four highest scoring teams are notified that they have won and are invited to attend the SuperQuest Institute. Other teams are awarded honorable mention as recommended by the committee. At the summer institute held in the first three weeks in July, students learn to use the IBM 3090-600Es and work with consultants who help the students to run their programs on the supercomputer.

The final phase of the contest, continued research, stretches from the September immediately after the summer institute to the following April. During this period, work stations are installed in the winners' schools, and the winners are encouraged to continue their research in their schools to make use of their new facilities and access to the supercomputer. The "best paper" and "best school" competitions, open to the winning students and schools, conclude the program in the spring.

Nature of Contest: The SuperQuest sponsors have several specific requirements for entries, as well as some general recommendations. The problems that the students attempt to solve must represent real science; participants should demonstrate a better solution to an existing problem than was previously available. The computer application software in its final version must be written in standard ANSI 77 FORTRAN. No projects involving human subjects or live vertebrate animals are eligible. The contest sponsors recommend problems that exploit the capabilities of the IBM 3090-600E supercomputer, but not simply as a supercalculator.

For Further Information: Write to SuperQuest, Cornell Theory Center, 265 Olin Hall, Ithaca, NY 14853. Telephone (607) 255-3985.

TECHNOLOGY STUDENT ASSOCIATION COMPETITIVE EVENTS

Background: While events were added at various times, the overall competition started in 1978. Some competitions receive funding from other sources, but all are administered by the Technology Student Association (formerly, the American Industrial Arts Student Association). Most of the 27 events have 2 levels of competition: Level I, for students in grades 7–9, and Level II, for students in grades 9–12. Only Level II events suitable for high school science and technology competitions are covered here.

Participants: This year there were 3,300 participants in the Technology Student Association Competitive Events. In order to compete at the national level, contestants must be members in good standing with the Technology Student Association and registered and in attendance at the national conference. Participants may enter a maximum of six events but may submit only one entry per event. For example, they may enter only one Metric 500 car. In team events, contestants must enter the event at the same level as their official school classification. For example, although ninth graders compete at both Levels I and II, they must enter at the same level as their school. Thus, any *chapter* (the Technology Students of America' in-school organization) with members at both levels may enter only at the level of its oldest member.

Awards: The first-, second-, and third-place national winners in each event receive trophies at the presentation ceremony. In addition, scholarships are awarded to the winners of the following Level II contests: Computer-Aided Draft/Design Engineering Problems, Electricity/Electronics, Graphic LOGO, and Outstanding Chapter. Finalists in all events are also recognized at the awards presentation ceremony and receive a certificate at that time. The number of finalists in a given event depends on the number of entries in that event.

When and Where and How Much: There are three tiers to the competition: within each school in March, at the state level in April, and at the national level in June. Students do not have to win or place to compete at the state and national levels—any contestant meeting the requirements may enter.

The national competitive events in June are held at the National Conference, in a different U.S. city each year. Contestants must pay their own travel expenses. Several of the events require that entries must be postmarked by May 1. The Technology Student Association has a strict dress code for participants in the national events, which is outlined in its *Competitive Events Guidelines*. There is a $50 fee for the National Conference registration, which allows participation in all conference activities, including competitive events.

Nature of Contests: Of the 27 officially approved events, 12 are science competitions suitable for high school students. Descriptions follow.

Bridge Building II. In this event, which is sponsored by Pitsco, Inc., two-member teams construct a model truss bridge out of balsa strips, according to assigned span and width specifications. The bridge dries a minimum of twelve hours; then judges test for design efficiency by applying weight until the bridge fails. Teams must provide the judges sketches of their designs before receiving their building materials and beginning construction. In the event of a tie, the sketches help determine the winner. Each chapter may enter only one team.

Computer-Aided Drafting/Design II. Sponsored by Versa Cad, Inc., Autodesk, Inc., and Modern School Supply, this contest is among the mail-in events. State advisors are responsible for submitting entries by the deadline. There may be only one entry per state in each of the categories, Mechanical Level I and II and Architectural Level I and II. These finalists travel to the national event in order to be tested in person on their computer-aided drafting and design skills, as applied to mechanical or architectural design and drafting. Contestants are given a problem and four hours in which to come up with a design solution. Judges base their evaluations on accuracy of the solution, placement of the views, dimensioning, use of computer-aided drafting and design functions, completeness, originality and creativity, use of linotype and pens, and drawing set-up.

Electricity/Electronics II. In this competition, contestants set up for judging a product using electricity or electronics they have designed and built within the past school year. This entry may be the result of individual or group work; however, if a chapter enters as a group, only one person represents the group in the interview. A one-page written description, a schematic drawing, and a parts list must accompany each entry. Judges award points according to these criteria: complexity of the entry, ingenuity, quality of work, schematic drawing, parts list, and written description. The finalists—those with the highest number of points—describe the project and demonstrate its use during a 10-minute interview, for which they may be awarded another possible 20 points. The highest scorers out of a possible 100 points are awarded first, second, and third place. Among winning devices was an amplifier and speaker system constructed for a portable tape player.

Engineering Problems II. Contestants submit a portfolio of sheets that demonstrate their knowledge of drafting and their ability to communicate technically. These drawings are judged on line quality, accuracy of details, dimensioning accuracy, and neatness—lettering uniformity and general appearance. The top 10 scorers are finalists, who then take an hour-long objective exam and a drawing test for another possible 60 points. First, second, and third place go to the top scorers. Only two individuals per chapter may enter.

Manufacturing Prototype II. Contestants submit for judging a prototype of a product they have made during the current school year. The products should have innovative features with relevant consumer applications. A "manufacturing scenario," (a document that explains the functions of the product and how it is made), and bill of materials must accompany the prototype. The product must include at least one material from three groups: wood, plastic, metal, and Earth products. Judges award points for the product (65 points), the bill of materials (10 points), and the manufacturing scenario (25 points). One winning project was a miniature air compressor.

Material Processes. Contestants mail in the development procedures, photographs, and working drawings of a project they have constructed during the current school year. Judges award points based on the complexity of the project, the working drawings, and the written development procedures with photographs. The top scorers are finalists, who bring their projects to the National Conference for further evaluation and a 15-minute interview. The national contestants are awarded points for the complexity of the project, ingenuity and design, quality of work, working drawings, written development procedures with photographs or explanation, and the interview. Past winning projects have included a computer table, a Plexiglas™ gum ball machine, and a lap desk. Each chapter may enter only once.

Metric 500 II. In this event, sponsored by Pitsco, Inc., contestants design, draw, and build a CO_2-powered dragster and bring it to the national conference. Judges evaluate the entries on the basis of design, speed, craft, and a drawing of the dragster. (Students must design, draw, build, and race the dragster within a rigid set of specifications.) After the judges rate these first three components, time trials are run to establish the dragster's official rank. The fastest 16 cars then compete in a double elimination bracket to determine final placing in the speed portion of the contest. The points earned in the design, drawing, construction, and race portions help to determine the winners, but the points are most heavily weighted in the race portion. Each chapter may enter only two contestants.

Radio Control Transportation. Three-member teams enter with a radio-controlled car they have built. The car is rated on how well it races, its appearance, a short technical report about it, and the modifications or changes made to it. Once the teams are judged on the car's appearance, the technical report, and modifications, each team member competes in one qualifying heat in the race portion of the event. Contestants try to complete the greatest number of laps within a set time. The twelve individuals with the most laps go on to the semifinals, and the top six semifinalists compete in the final race. Points are awarded for the appearance, technical report, modifications, and race portions of the events, with the most points awarded in the race portion. Each chapter may enter one team.

Research Paper II. Students mail in papers by early May to demonstrate their ability to select a topic, research it, and write up their findings according to an accepted format. The paper must be no longer than 10 pages; the topic may be related to any aspect of technology education or industrial arts. Each chapter may enter two contestants. The judges evaluate each entry based on the paper's organization, how well the research is supported by evidence, grammar and spelling, and topic originality. Winning papers have discussed telecommunications and satellite transmissions.

Technical Report Writing II. This event is designed to give members the chance to demonstrate their ability to consolidate information and prepare a technical report. Contestants have one hour in which to read and analyze technical articles and write a comprehensive report summarizing key elements. The reports are judged on content, organization, interest, and neatness. Each chapter may enter two members.

Technology Bowl II. This event gives students the opportunity to demonstrate their knowledge in five areas: Technology Student Association, construction, manufacturing, transportation/power and energy, and communication. Each chapter may enter one three-member team. To qualify for the oral team competition, individuals must first complete a written objective exam. The top-scoring eight teams compete in an oral question-and-answer, head-to-head elimination round, so that in the next round there are four teams; in the next, two; the winner of which is champion of the oral exam. Those who place first, second, and third in both the written and oral exam portions receive awards.

Technology Process Display II. This event is for Technology Student Association chapter teams and gives chapters the opportunity to demonstrate their knowledge of a technological process they have researched by creating a display of that process. The exhibit must depict some technological or industrial operation; judges encourage the display of applications of new technologies that solve technical problems. A written description of the process must accompany the exhibit. Exhibits are rated on logic and sequence of the presentation, originality, subject coverage, interest and appeal, and quality of work. One past winning display featured the process of manufacturing radial tires.

For Further Information: Write to Rosanne T. White, Executive Director, Technology Student Association, 1914 Association Dr., Reston, VA 22091. Telephone (703) 860-9000 or (703) 620-1060.

VOCATIONAL INDUSTRIAL CLUBS OF AMERICA (VICA) U.S. SKILL OLYMPICS

Background: The U.S. Skill Olympics were first held in 1968 by the Vocational Industrial Clubs of America, but the official name was established only in 1973. The competitions are funded by over 400 organizations; among them, corporations, trade associations, and labor unions. Administered by Vocational Industrial Clubs of America, the Skill Olympics today comprise 43 different contests—a forum for vocational students to demonstrate their technical and leadership skills. The Skill Olympics are designed not only to reward students, but also to involve industry in assessing student performance and to help make vocational education relevant to industry's future needs. To this end, the Skill Olympics' nearly 1,000 judges and organizers come "from the ranks of labor and management."

Participants: There are approximately 2,800 participants at the national level competition. To enter nationally, they must first compete at local and state levels where the Skill Olympics are open to students who are active, dues-paying members of the Vocational Industrial Clubs of America. Participants in the national contests must be first-place winners at the state level and must be so certified by the state association. Secondary-level contestants must be enrolled in an occupational program that pertains to the contest they wish to enter and that has been approved by the state department of education. In addition, they must be earning credit toward high school graduation during the school year immediately preceding the Vocational Industrial Clubs of America National Leadership Conference. There are additional eligibility requirements for some contests.

Awards: Gold, silver, and bronze medallions are awarded for first, second, and third place. Some contests also award tools and equipment.

When and Where and How Much: Winners of the state Skill Olympics who wish to compete nationally must submit an official National Leadership Conference card within 10 days of the finish of state Skill Olympics, usually in early March. Contestants *must* attend the orientation meeting that is held at the national conference site prior to the contests. Contestants travel to the conference at their own expense. For information on local and state Skills Olympics, contact the Vocational Industrial Clubs of America U.S. Skill Olympics national headquarters and ask for the name and telephone number of the state association officer.

Nature of Contests: Of the 43 contests in the Skill Olympics, 6 are appropriate for high school students. Descriptions follow.

Automotive Service Technology. The major areas of the contest are inspection, service, and repair of automotive systems, including cooling, heating, and air conditioning; various electrical systems; tires and wheels; steering and suspension; brake systems; engine and component parts; exhaust systems; fuel systems; and drive train. There is also a written test administered by the National Institute for Automotive Service Excellence. Contestants are judged on their skill, accuracy, quality of work, safety practices, and speed in comparison with the factory specifications.

Aviation Maintenance Technology. The contest for secondary participants covers general aviation skills as classified by the Federal Aviation Administration. Contestants perform from 8 to 15 assigned operations, which are further broken down into specific criteria. For example, areas covered might include service, such as repair of a sheet metal structure, use of manuals, use of tools, or testing engine operation.

Computer-Aided Drafting. Contestants demonstrate their ability to perform basic machine drafting jobs with a computer-aided drafting work station within a set time. Computer-aided drafting skills include establishing and setting system parameters; using x, y, and z coordinate systems; using layers/levels, colors/pens, and line types; using snap features; making dimensional commands; calculating sizes; macro programming; hatching; using isometrics; three-dimensional modeling; digitizing; plotting; or using MS–DOS fundamentals. The contest focuses on problem-solving techniques, using visualization, geometric construction, and dimensioning skills, and emphasizes accuracy, speed, scaling, the use of colors, layers, and plotting. Contestants are briefed on any additional criteria during the precontest meeting.

Electronic Products Servicing. Contestants demonstrate their skills in these major contest areas: customer relations, diagnosis and service of electronic systems, color television circuits, audio systems and turntables; testing and troubleshooting for audio amplifiers, AM/FM tuners, magnetic tape systems, and speaker systems; analyzing digital systems and circuits; microprocessing functional analysis and software; service of VCRs and knowledge of VCR fundamentals; and troubleshooting for and service of VHS and BETA video recording equipment. Winners are determined on the basis of diagnostic procedure, speed, standard industry procedures, accuracy of adjustments, and correct component replacement.

Electronics Technology. Contestants demonstrate their ability to perform assignments within a set time in the following areas: technical recording and reporting, DC circuits, AC circuits, solid state devices, analog circuits, digital devices, microprocessing, and lab practices. Judges test the completed projects for quality of work and operating specifications.

Precision Machining. Contestants demonstrate their skills in the following areas: mathematical calculations; designing and planning machine work; performing metalwork operations and benchwork; operating drill presses, grinding machines, lathes, milling machines, and power saws; and maintaining machines, tools, and shop facilities.

For Further Information: Write to Harold E. Lewis, Vocational Industrial Clubs of America U.S. Skill Olympics, P.O. Box 3000, Leesburg, VA 22075. Telephone (703) 777-8810.

WESTINGHOUSE SCIENCE TALENT SEARCH

Background: The Westinghouse Science Talent Search was started in 1942 by Westinghouse Electric Corporation. The Science Search is administered by Science Service, a Washington-based nonprofit organization engaged in furthering public understanding of science, which also sponsors the International Science and Engineering Fair. (See page 137.) Since the Search started, over 100,000 students have completed independent research projects and submitted written reports. The victors in the early Searches were often refugees—frequently German Jews whose families had fled Hitler. Many of today's winners are also first-generation immigrants—now mostly Asian Americans.

Certain schools in certain areas tend to produce the most finalists: By far the highest number come from New York. Almost a third of the 1,920 winners went to high school in New York state, some of them attending selective institutions specializing in science (like the Bronx High School of Science, which has produced 113 winners—more than any other). Others attended both private and public schools, some with special programs or courses designed especially for Westinghouse entrants. Many participants take independent study courses that allow them to work on their projects; some receive other incentives to join the Search.

Arguably the most prestigious of high school science competitions, the Search is the ancestor of national and international science competitions for high school students, although science fairs began to be held in many schools earlier. All major contests, from national ones like the Junior Science and Humanities Symposia or Duracell NSTA to the Olympiads held among students from many nations, are in a sense descendants of the Search. Among the Search's past winners are 5 Nobel Laureates,* 2 Fields Medalists, 8 MacArthur Fellows, and 26 members of the National Academy of Sciences.

Participants: All are high school seniors: About 15,000 applications are requested; some 1,500 students enter; 300 make the semifinalist—or honors—group; 40 win.

Awards: Collectively, winners receive a total of $140,000: 1 $20,000 scholarship; 2, $15,000; 3, $10,000; 4, $7,500; and 30, $1,000.

When and Where and How Much: Participants must submit a report on an independent research project and a personal data form in mid-December. Judges first select the semifinalist, or honors, group and then pick the top 40 finalists. Both groups are announced in January. Westinghouse brings the finalists—all expenses paid—to Washington, D.C., for further evaluation. The ten top scholarship winners are selected on the basis of interviews and their research.

Nature of Contest: Students submit a report on an independent research project in the physical sciences, behavioral and social sciences, engineering, mathematics, or biological sciences.

They spend from six to nine months to several years on their research. President of Science Service, E. G. Sherburne, Jr., offers these "words of wisdom" to prospective entrants: "Research takes time. You can't start a winning project at Thanksgiving and hope to win in December," he warns, adding that many projects take five years of work or even longer. For example, one winning entry on the cuticle of the fiddler crab was started when the future finalist was a small child fascinated by what he found in a tidal pool at the seashore. The boy's interest in fiddler crabs deepened as he entered biology and then chemistry classes and culminated in his Westinghouse project some ten years later.

Teacher/advisors involved in the contest characterize some of the students' efforts as graduate-level work. One finalist's research

*Two of whom, Sheldon Glashow and Roald Hoffmann, responded to NSTA's survey and are quoted below (See parts III and IV). Glashow's response is reproduced on pages 186-187.

involved constructing a biomechanical model of the requirements for takeoff in pterosaurs and testing the model by using measurements of fossils and comparable living fliers such as albatrosses. Her finding— that pterosaurs were capable of erect, bipedal running take-offs with lift and balance probably provided by their wings being held up and swept back—appears to help resolve a century-old controversy over whether pterosaurs were bipedal.

Students and their advisors fill out a *Personal Data Blank* designed to show evidence of characteristics and habits the judges believe are important to becoming a creative scientist. Questions reveal the students' work habits, extra-curricular interests and activities, inventiveness, curiosity, initiative, and scientific attitude, as well as family background and academic standing.

Students' Comments:

1. One student writes that it is important "to remember [that] the project is only a part of what the judges are looking for. For example,…assuming your project has the required degree of originality, it will not be referred to again in the interview part of the competition…in Washington. There, the judges will be looking for talent in science: That is, how much have you managed to learn in science? This is not necessarily…encouraging [to] students in science, seeing that the quality of science teaching can be extremely variable, but there is probably no other way a science award can be determined."

2. Another, whose project took three years, describes an arduous schedule: "I collected and processed my specimens every summer during vacation and then analyzed the data in the fall, ultimately presenting my work at fairs which began in the winter. I'd work year-round, three days a week on average during the summers, weekends when school started. Holidays were exclusively set aside for my experiments."

3. A third remembers, with gratitude, help received. "The three teachers who directed the science camp were always willing to answer questions, help brainstorm ideas for the projects, and accompany students to the state library for background research. On separate occasions, two teachers accompanied me to Hebgen Lake, driving four hours each way to collect toxic water samples. The teachers also worked closely with me throughout the school year when I wrote my paper. In addition to the aid of my teachers, I had the support of the Water Quality Bureau, which provided much information about toxic blue-green algae and

about Daphnia. This bureau also supplied food for my Daphnia cultures. Finally, the state Diagnostic Laboratory in Bozeman, Montana, conducted tests of my water samples using their mouse bioassay, giving me a basis by which to compare the results of my own bioassay."

4. A fourth is wry: "We're in sad shape if I represent the 'best science talent' in America. Let's work at it."

Teachers/Mentors' Comments:

1. One teacher writes, "I provided initial instruction (in a research in biology course) in selected laboratory techniques and suggested possible research problems. I also provided transportation to the university science library, met once weekly to go over [the student's] past week's progress and plan the next week and performed a test that [she could not do] (for safety reasons). I also praised her a lot, which she deserved."

2. Another is also positive: "Students get an excellent idea of high-level research. They realize the complexity of this level through the detailed data form and written report. If they are recommended students or state-level winners—and especially if they are in the top 40 students—they have an experience of having achieved something few others have ever achieved."

3. A third adds that, while the concerned teacher can help boost discouraged students who "lose enthusiasm" as the project proceeds, often s/he can also provide access to "someone knowledgeable in the field," who can give "a pat on the back." The teacher continues, "If at all possible, get them in contact with someone who is truly expert in what they are doing so they feel they are on the forefront of knowledge."

Judge's Comments: Nobelist Glenn T. Seaborg,* who has judged the Science Talent Search for 25 years, says that he looks for "breadth of knowledge and understanding of science as determined by an oral interviewer." He believes that the practice of having students present and interviewed individually is the "best method" of presentation.

For Further Information: Write to Science Service, 1719 N Street, N.W., Washington, DC 20036. Telephone (202) 785-2255.

*Seaborg's survey is reproduced on pages 184-185.

SCIENCE INTERNATIONAL

INTERNATIONAL COMPUTER
ANIMATION COMPETITION

Background: In 1985, to recognize excellence in computer animation, the National Computer Graphics Association began sponsoring and funding the International Computer Animation Competition.

Participants: Of the 115 contestants last year, only one was still in high school—the others were postsecondary academics or professionals in broadcast–computer graphics; television commercials; corporate presentations; technology and computer graphics research; and/or short films, videos, and theatrical-motion computer graphics. The Association, however, encourages high school students to enter.

Awards: As appropriate, in the categories just listed, first-place winners receive trophies and tickets to the awards dinner held in late March in Anaheim, California. Holders of second- and third-place prizes and honorable mentions receive certificates and dinner tickets. Winning videos by first-, second-, and third-place winners in each category will be shown at this dinner; the Association will also publicize winning videos in other ways. No transportation is provided.

When and Where and How Much: Entry fees for academics (includes high school students) are $25 for each nonprofessional category and must be sent by late November to the competition coordinator of the Computer Animation Competition (address below).

Nature of Contest: Eligible are no more than five-minute long three-quarter or one-half inch VHS videotape entries, produced on digital (not analog) computers since December of the preceding year. Contestants must also submit at least one 35 mm slide of selected frames. To enter the student/faculty category, entrants must be certified as such by a school official.

For Further Information: Write or call Tanya Bosse, Computer Animation Competition, National Computer Graphics Association, 2722 Merrilee Drive, Suite 200, Fairfax, VA 22031. Telephone (703) 698-9600.

INTERNATIONAL COMPUTER PROBLEM-SOLVING CONTEST

Background: Started in 1977 as a contest based at the University of Wisconsin—Parkside and held in conjunction with an annual microcomputer fair, the International Computer Problem-Solving Contest went nationwide in 1981 and worldwide shortly thereafter. From 1981 to 1988, the contest was sponsored by the University of Wisconsin—Parkside; it currently sponsors itself, with financial support from "USENIX," the technical association of users of UNIX (a computer operating system).

Participants: Any school in the United States or abroad may field a "team," which can include a single member or up to 3 players under age 18 at one of the following levels and languages: elementary (grades 4–6, LOGO and BASIC); junior (7–9, LOGO and BASIC); and senior (10–12, LOGO, BASIC, and Pascal). The contest currently serves approximately 5,000 teams from 45 states and 12 foreign countries.

Awards: All teams that solve the five problems posed are given a certificate of achievement. The top-ranked team in each age and computer language division also receives a plaque with winners' names engraved upon it. The programs, written by the top teams, are published in *The Computing Teacher* and *LOGO Exchange*. In addition, organizers of contests often provide other awards to winning teams.

When and Where and How Much: The Contest is held on the last Saturday in April at local sites throughout the world. There is a $20 charge for entering the contest, and, because the competitions are held locally but publicized worldwide, no travel is necessary.

Nature of Contest: The International Computer Problem-Solving Contest annually challenges teams of one to three students at three levels to solve five computer problems using any programming language and on any computer system. The Contest combines the art of problem solving with skill at computer programming. The five categories of problems are computation, simulation, graphic patterns, words, and mind benders.

In response to a request from a teacher or another interested party, the International Computer Problem-Solving Contest furnishes all materials necessary to conduct a local contest: problems, solutions, score cards, rules, judges' information, and the like. Contest headquarters will also provide sample problems and solutions to help teachers and students to practice.

The average number of problems solved by individual teams is two out of five. Contests are judged locally; the programs of teams solving all five problems correctly are sent to the national contest headquarters for ranking with the best results worldwide.

For Further Information: Write to Donald T. Piele, International Computer Problem-Solving Contest, P. O. Box 085664, Racine, WI 53408. Telephone: (414) 634-0868. Or read *The Computing Teacher* (especially the October, November, and December issues).

INTERNATIONAL SCIENCE AND ENGINEERING FAIR

Background: The International Science and Engineering Fair was started by Science Service in 1950 and is funded by affiliate fees, the host city, Science Service, and award sponsors (professional societies, the government, and corporations). Like the Westinghouse Science Talent Search, which is also administered by Science Service, the Fair encourages student interest in scientific research.

Participants: Over one million students in ninth to twelfth grades from the United States, several of its territories, and 10 other countries each year feed into the 380 affiliated fairs. No more than 2 students—almost 800—from each affiliate win the right to represent their fair at the International Science and Engineering Fair.

Awards: All finalists are winners and receive a rainbow-ribboned silver medal. In addition, Science Service and the Fair's special awards organization offer more than 550 awards. Science Service sponsors first-, second-, third-, and fourth-place cash Grand Awards in each of the research categories ($500–$100) and sends two finalists to the Nobel Prize ceremonies in Stockholm, Sweden.

When and Where and How Much: Affiliate fairs (most of which cover at least two counties) must be held by mid-April; Science Service must receive finalists' paperwork by late April. The Fair is held in the second week of May, in a different U.S. city each year. Affiliates pay round-trip travel, living, and project-shipping costs for up to two finalists and one adult escort to the Fair.

Nature of Contest: Students may submit research projects in these categories: behavioral and social sciences, biochemistry, botany, chemistry, computer science, Earth and space sciences, engineering, environmental sciences, mathematics, medicine and health, microbiology, physics, and zoology. Projects are evaluated by local judges whose number sometimes exceeds that of participants.

The path to the Science and Engineering International Fair must be through victory in an affiliate fair.

Finalists, their affiliated fairs and boards, their teacher/mentors, and other supervising adults (such as animal care supervisor, supervising scientist, etc.) must observe numerous rules, requirements, and deadlines provided in a 32-page booklet of regulations and forms. In addition, adult supervisors must accompany finalists to the Fair, and finalists must be present for judges' interviews. Science Service publishes a booklet listing the finalists' research projects and a book of their abstracts. The contest receives press coverage nationwide.

For Further Information: Write to Science Service, 1719 N Street, N.W., Washington, DC 20036. Telephone (202) 785-2255.

THOMAS EDISON/MAX MCGRAW SCHOLARSHIP PROGRAM

Background: This program, initiated in 1979 by the Thomas Edison Foundation and the Max McGraw Foundation, was interrupted in 1987 and restarted in 1989 under the sponsorship of the Max McGraw Foundation. The National Science Supervisors Association coordinates the program. The purpose of the program is to recognize, by awarding ten scholarships, students who most nearly demonstrate the genius of scientist and inventor Thomas Edison and philanthropist and inventor Max McGraw.

Participants: The contest is open to students in grades 7–12 in public, private, and parochial schools in the United States, Canada, and other participating nations. The contest has two divisions, junior and senior, for students in grades 7–9 and 10–12, respectively.

Awards: The winners of the two divisions each receive a $5,000 scholarship. The four remaining finalists in the senior division receive $1,000 each and the four other junior division finalists receive $500 each. Senior division winners' scholarships are paid directly to their chosen college after the students graduate from high school. Junior division winners receive their cash awards directly. Certificates of recognition are awarded to the ten finalists' teachers/sponsors and their schools.

When and Where and How Much: By late November, students submit their proposals and a teacher or sponsor's letter of recommendation that indicates how the student's proposal exemplifies the creativity and ingenuity of Edison and McGraw. The typewritten or word-processed proposals should not exceed 1,000 words or five single-spaced pages, written in standard English. Twenty-five semi-finalists are notified by February 1. Then, in the spring, the judges select ten finalists, who travel to the judging site to present their projects to the team of judges. Contest sponsors fund all travel expenses.

Nature of the Contest: Contest participants submit a proposal on a completed experiment or a projected idea that has practical application in the fields of science and/or engineering. Judges look for creativity and ingenuity but also for practicality.

For Further Information: Write to Kenneth Roy, National Science Supervisors Association Leadership Institute for Science Education Center, Copernicus Hall, Room 227, Central Connecticut State University, 1615 Stanley St., New Britain, CT 06050. Telephone (203) 827-7981.

UNITED STATES NATIONAL CHEMISTRY OLYMPIAD

Background: The International Chemistry Olympiad originated in 1968 with participants from Czechoslovakia, Poland, and Hungary. Other Eastern Bloc nations soon joined the competition; Western European countries began to enter in 1974. In the same year, the United States fielded its first team, competing in successive years in the Federal Republic of Germany, Czechoslovakia, the Netherlands, Hungary, Finland, and the German Democratic Republic.

The competition is administered and funded by the American Chemical Society with the participation of its local sections.

Participants: All participants must be U.S. citizens under 21 years of age and affiliated with one of 183 local sections of the American Chemical Society. Leaders and advisors of these sections begin to sift through the 140,000 hopefuls who would like to compete in the Chemistry Olympiad. The top 700, comprising students selected according to local options, take another test—this one written and graded nationally but administered locally. It is set up much like an advanced-placement exam in chemistry, with two hours of multiple-choice questions and two hours of essay writing. The top 20 students on this test, all considered as winning members of the Chemistry Olympiad team, spend about two weeks in midsummer in a chemistry study camp in Colorado. Four of these students go abroad to participate in the International Chemistry Olympiad.

Awards: At the national level, the top 20 contestants win a two-week, expenses-paid trip to the chemistry study camp in Colorado. The top four travel, again, expenses-paid, to the city where the International Chemistry Olympiad is held, where they are eligible to win bronze, silver, and gold medals. All participants receive certificates.

When and Where and How Much: From the original 140,000 contenders, the local sections choose before mid-March about 700 on the basis of participating students' performances on locally or nationally developed exams, in chemistry competitions among students ("chemathons"), in laboratory work, by teacher recommendations, in science fairs, or through a combination of the five. The next cut to 20 occurs through American Chemical Society exams taken locally about the middle of April but graded nationally. The 20 top scorers go on to a 10-day study camp in Colorado, where they work intensively until the four best are selected to go for a week to the International Chemistry Olympiad abroad. All expenses are paid by the American Chemical Society and the U.S. Air Force Academy.

Nature of Contest: Although the U.S. Chemistry Olympiad is essentially a contest, its aim is not merely to win. It also strives to

- stimulate young people to achieve excellence in chemistry
- recognize outstanding chemistry students and...encourage their additional learning

- recognize the achievements of students' teachers and the importance of their school environment
- promote contact between the local sections, schools, and professional chemists
- foster cross-cultural experiences

Although students go to the Olympiad as part of a team, individuals win (or don't win) medals on their own. All contestants at the Olympiad take a test in two parts—a five-hour theoretical written exam and a five-hour practical lab exam.

Student's Comments: One student, after thanking his teacher for encouraging "but not pushing" and for driving him and others to a distant city to take the locally administered tests, makes this observation: "I found it interesting (but not at all surprising) that so many of us at the camp were musicians...I play cello and compose music, and I'm very interested in electronic music."

Teachers/Mentors' Comments:

1. Concerned with teachers' serious time constraints, one, who spends about 10 hours a month helping students, writes, "Science teachers in general do not have the time to devote to competition activities with various preparations, lab preparations, and the like." The teacher also notes, however, that motivated students "have their own drive."

2. Another teacher, among whose students are winners in the Science and Chemistry Olympiads as well as in school-based contests, also comments on the time problem. She notes that science contests give the "'non-jock' students some recognition that they don't normally get. It may also help keep students in high school who might have gone on to college in their senior year." Pointing out that students had to raise $10,000 to go to the national Science Olympiad, the teacher suggests further, "The school districts are going to have to treat these competitions like sports—football, basketball, etc.—and offer similar financial support."

For Further Information: Write Martha Turckes, Education Division, The American Chemical Society, 1155 16th Street, N.W., Washington, DC 20036. (202) 872-4382.

UNITED STATES PHYSICS TEAM FOR THE INTERNATIONAL PHYSICS OLYMPIAD

Background: The International Physics Olympiad began in Eastern Europe in 1967 and gradually expanded to become a worldwide competition rewarding and recognizing superior precollege physics students. The United States fielded a team in 1986. U.S. students competed in 1986 in England; in 1987 in the Democratic Republic of Germany; in 1988 in Austria; and in 1989 in Poland.

Started by Jack M. Wilson, Chief Executive Officer of the American Association of Physics Teachers, the team continues under that organization's administration. Fund-raising, however, is run by the American Institute of Physics, which has solicited generous contributions and donations from 13 scientific associations, 24 business corporations (including 6 publishers), and students and administrators from 2 universities.

Participants: The Association sent applications to 5,700 high schools and scattered other sources, each of which may nominate candidates. Most nominate only one; others, none; some, several. It is estimated that 10,000 students nationwide receive information about the contest. Entrants, who must be under 19 and in high school full time, compete in the early rounds to be part of the team going to the Physics Olympiad. From the local selection process emerge about 400 students who take a national exam; the top 70 scorers in this group are eligible to take a test similar to that given at the International Olympiad. From this group are chosen the top 20, all of whom are honored as members of the U.S. Physics Team for the International Olympiad but only 5 of whom are chosen to travel to the international competition.

Awards: At the national level, the top 20 physics students participate, all expenses paid, in a 10-day training camp held in late May at the University of Maryland, College Park. The top five travel, again expenses paid, in July to the International Physics Olympiad, held in Europe, where they are eligible to win bronze, silver, and gold medals.

When and Where and How Much: About 500 students are initially selected after schools receive mailings in November via locally determined options that include tests (including samples furnished by the Association), competitions, teacher recommendations, and/or grade point averages. Nominations, accompanied by a $20 application fee, are due in mid-January. The first-round exam, made up of both multiple-choice and open-response questions, is held in individual schools early in February for all those nominated. Teachers in the schools grade the exams.

The semifinals, made up of 5 open-response questions and graded by the academic advisors, are given to the 60 top scorers in the first round at their local schools early in March. Twenty of these are chosen as finalists and go to the study camp in May, where they prepare for the Olympiad held in Europe.

Nature of Contest: The 10-day summer study camp offers the 20 members of the team intensive participative training in many facets of physics. Its philosophy, adapted from that of successful athletic coaches, is first to bolster students' sense of self-worth and convince them that they could win the Olympiad, and, second, to teach physics. Time for relaxation is built into the program.

Although students participate in the Olympiad as part of a team, individuals win (or don't win) medals as a result of scores on a two-part examination (each half requiring five hours) comprising three theoretical problems and two experimental ones. Members of the U.S. team in 1986 won three bronze medals, compiling the best record of any team competing in the Physics Olympiad for the first time. In 1989, Steven S. Gubser* of Colorado came in first, scoring 46 points out of a possible 50 and thereby placing above 150 competitors from 30 countries.

For Further Information: Write to the International Physics Olympiad, American Association of Physics Teachers, 5112 Berwyn Road, College Park, MD 20740. Telephone (301) 345-4200.

*Gubser's questionnaire is reproduced on pages 180-181.

YOUNG PHYSICISTS' TOURNAMENT

Background: Held in the USSR since 1979, the Young Physicists' Tournament is a collective competition that allows high school students to demonstrate their ability to solve complicated physics problems, to justify their positions against their peers' questions and objections, to present solutions to these problems in convincing forms, and to participate in scientific discussions, which are called "physicists' fights." The Tournament has become a traditional form of work for students aged 14–17, held at the physics department of Moscow State University.

Participants: In 1989, invitations were sent to high schools in Australia, Bulgaria, the People's Republic of China, Czechoslovakia, Finland, the Federal Republic of Germany, the German Democratic Republic, Hungary, Italy, Kuwait, Japan, Yugoslavia, the Netherlands, Poland, Romania, Great Britain, and Spain. The United States will be invited in future years; however, any school—whether invited or not—that wishes to send a team may enter.

Awards: No formal prizes are presented.

When and Where and How Much: The first stage, which takes place in September and November, is a collective competition conducted by mail and open to any school team that wishes to participate. Usually the teams face a list of 17 rather complicated physics problems or tasks. Although high school students' original studies are desirable, the problems are multilevel in nature and high school teams are often joined by postsecondary students, scientists, or even whole scientific institutes. Stage two, which consists of scientific discussions (or "fights") takes place in December and January in various countries (in the USSR in 1989; in Czechoslovakia in 1990). In stage three, held in late March in Moscow, the results of the Tournament are discussed and a winner is declared. While contestants' schools must pay participating students' travel expenses, the Young Physicists' Tournament charges no entrance fees.

Nature of Contest: In the first stage of the contest, students face multilevel problems in outline form offering freedom to use creative initiative to specify the problem, to choose ways to solve it, and to decide in how much detail to pursue it. This framework, along with the wide levels of expertise among the participants, can lead to noteworthy results.

The second level consists of a series of timed contests where participants defend their solutions in front of rival contestants, an interested audience, and an impartial jury. First, a questioner asks a team representative, a "reporter," to solve any one of the original 17 problems. The reporter has five to eight minutes to present the essence of the solution, drawing the audience's attention to main ideas and conclusions. The original questioner has four minutes to question the presentation, pinpointing errors and inaccuracies. Then, a reviewer offers a one-minute commentary on the two presentations. Finally, a jury—made up of leading scientists, professors, teachers, postgraduates, and students from Moscow State University—makes its judgment.

For Further Information: Write to E. Unosov, c/o Sergei Krotov, *QUANTUM* Bureau, Gorcy St., Moscow 103006, USSR. Telephone (011) 7 095 251-8242; FAX: (011) 7 095 251-5557.

Math Competitions—
Nationwide and Worldwide

MATHEMATICS COMPETITIONS: AN OVERVIEW

Edward D. Lozansky

In many competitions in science and math, as in sports, the highest award is the Olympic medal. Only a few participants will get gold, silver, or bronze medals, but merely participating in the main contest, preparatory sessions, or initial local or national competitions can be a rewarding, stimulating, and inspiring intellectual experience.

Each summer, the United States sends teams to three different international academic olympiads for high school students—mathematics, physics, and chemistry. At the present time, plans are being made to add an olympiad in space research. There are other international competitions as well (see pages 134-145).

Competing for a place on one of those teams can be one of the most exciting events in the life of a high school student. Of course, much depends on the enthusiasm of local high school teachers.*

A professor of mathematics and physics, Edward D. Lozansky is also coordinator for many NSTA international programs.

---

*MATHCOUNTS, a contest for 7th and 8th graders, is a cooperative project of several corporations, associations, and government agencies.

MATH NATIONAL

AMERICAN HIGH SCHOOL
MATHEMATICS EXAMINATION

The American High School Mathematics Examination (AHSME) is one of four challenging mathematics contests written and administered each year by the Committee on American Mathematics Competitions. This Committee also administers the American Junior High School Mathematics Examination, the American Invitational Mathematics Examination, and the USA Mathematical Olympiad, as well as oversees U.S. participation in the International Mathematical Olympiad. The purpose of the American High School Mathematics Examination is to identify students with an interest in and talent for mathematical problem solving.

Background: The American High School Mathematics Examination was organized in 1949 and marks its 41st anniversary with the 1990–1991 program. The American High School Mathematics Examination is sponsored by eight organizations: the Mathematics Association of America, the Society of Actuaries, Mu Alpha Theta, the National Council of Teachers of Mathematics, the Casualty Actuarial Society, the American Statistical Association, the American Mathematical Association of Two Year Colleges, and the American Mathematical Society.

Participants: Each year, the examination is given to over 350,000 students attending some 6,600 schools. All secondary and presecondary school students in the United States and Canada are eligible to participate through their schools. Although the competition is primarily intramural, regional results make some comparison at that level possible. The examination, however, has no established norms and is not comparable to either classroom or placement tests. Students who achieve a score of at least 100 on the American High School Mathematics Examination are invited to compete in the Committee on American Mathematics' American Invitational Mathematics Examination.

Awards: A pin or medal is awarded to the top students at each school. Regional and national awards include pins, plaques, and certificates. Some state scholarships for contest winners are available from regional examination coordinators.

When and Where and How Much: The American High School Mathematics Examination test is administered in the United States, at U.S. schools abroad, and through American embassies and military bases overseas. The examination is given in high schools in late February or early March. The schools are responsible for payment of registration fees ($15 per school), with a minimum exam order fixed in multiples of 10 for each school at a cost of 60 cents an exam. The registration deadline is in December.

Nature of Contest: The AHSME consists of 30 multiple-choice questions. Students are given 90 minutes to complete the test. A minimum score of 100 is required for inclusion in the individual honor roll. Scores of 280 and 350 are required for the school merit roll and the school honor roll respectively. The exam focuses on problem solving and covers all pre-calculus mathematics. The test, which must be administered on a day designated in advance, is machine scored.

For Further Information: Write to Walter E. Mientka, Executive Director, American Mathematics Competitions, Department of Mathematics and Statistics, University of Nebraska, Lincoln, NE 68588-0322. Telephone (402) 472-2257.

CONTINENTAL MATHEMATICS LEAGUE

Background: The Continental Mathematics League is a contest for elementary, junior high school, and high school students. Through-out the school year, schools that register with the Continental Mathematics League are provided tests to administer to students.

Awards: Each team receives five certificates and two medals. Some national and regional awards are also given.

Participants: Participants are students in grades two to three (one contest), and grades four to nine (four contests), one contest for calculus students, and three computer contests, one of them calling for Pascal. Any number of students may participate.

When and Where and How Much: Participants register in October. The fee is around $45.

Nature of the Contest: The tests focus on problem solving. Students may take tests at or above grade level. Tests for each division are given on or near dates specified in advance. Final results are sent to all participating schools.

For Further Information: Write to Joseph Quartararo, President, Continental Mathematics League, P.O. Box 477, Hauppauge, NY 11788. Telephone (516) 265-4792.

NATIONAL MATHEMATICS LEAGUE

Background: The National Mathematics League annually offers middle, junior high, and high school students 35 chances to compete in exams offered in 7 mathematical categories. Started in 1983 by mathematics coaches at Coral Spring High School, Broward County, Florida, the National Mathematics League was organized without sponsor funding.

Participants: Students in grades 6 to 12 are eligible to participate. Over 1,100 schools enter the contest each year. Participating students must be enrolled in courses roughly corresponding to the competition's seven divisions ("sixth grade," pre-algebra, algebra I, algebra II, geometry, pre-calculus, and calculator competitions for high school students). The exceptions to this rule are students in the pre-calculus and calculator competitions.

Awards: Ribbons for first- and second-place winners are awarded to the schools according to the contest divisions. An engraved plaque is awarded to schools amassing the highest cumulative scores in each division.

Top-scoring individuals are awarded rosette ribbons. Starting in 1990, the National Mathematics League plans to institute a special sweepstakes award for the top 10 schools.

When and Where and How Much: The entry deadline in the National Mathematics League competition is in late October. Registration arrangements are made through the schools, which pay $40 per contest division. Five separate contests in the competition's seven divisions are scheduled on different dates between January and April.

Nature of Contest: After each contest, the National Mathematics League newsletter provides schools with cumulative results of the event just completed as well as the results of the preceding contest. A sponsor, appointed by the school, receives all National Mathematics League mail and is also responsible for administering, proctoring, and scoring the tests. Each of the five contests, scheduled separately throughout the winter and early spring, involves a set of six open-answer problems to be completed within 30 minutes. The school's sponsor scores the tests using the National Mathematics League's answer key, which includes acceptable equivalent answers. The team score for each division consists of the sum of the five highest individual scores.

For Further Information: Write to the National Mathematics League, Box 9459, Coral Springs, FL 33075. Telephone (305) 344-8980.

MATH INTERNATIONAL

INTERNATIONAL
MATHEMATICAL OLYMPIAD

Background: Since 1959, when Romania initiated the International Mathematical Olympiad, the world's most mathematically accomplished high school students have gathered each year to participate in the competition. The United States entered the International Mathematical Olympiad for the first time in 1974, when it scored second among the participating countries. Since then, the U.S. team has generally placed among the top 10 competitors and was first in 1977 and 1981, tying with the USSR for first place in 1986.

Participants: Under the U.S. training program, 20 to 24 students who are top scorers in the USA Mathematical Olympiad are chosen as International Mathematical Olympiad candidates. Selection to participate in the USA Mathematical Olympiad is based on an indexed score of both the American High School Mathematics Examination and the American Invitational Mathematics Examination with the goal of obtaining 150 national-level participants. The top participants in the USA Mathematical Olympiad become candidates for International Mathematical Olympiad training, but there is no guarantee that a national winner will have a position on the international team.

Awards: The host country presents gold, silver, and bronze medals and other awards to the winners.

When and Where and How Much: The training session for the U.S. team in the International Mathematical Olympiad's is held in alternate years at the U.S. Military Academy (West Point) and the U.S. Naval Academy (Annapolis). Participants in the International Olympiad are selected during the first week of training based upon a battery of International Olympiad-type exams, as well as the candidates' score on the USA Mathematical Olympiad. The

Olympiad itself occurs each year in July in a designated country. Recent host countries have been Cuba, Australia, and West Germany. The training session and travel to the Olympiad are funded by the U.S. Office of Naval Research, the Army Research Office, and the Matilda R. Wilson Foundation. Expenses incurred while students participate are borne by the host country.

Nature of the Contest: The training sessions are highly intensive. Because sessions are held on military bases, participants must agree to comply with both the study rules of the program and the living rules of the base. The USA Mathematical Olympiad, like the International Mathematical Olympiad, is based on non-calculus mathematics. Students work on extremely difficult problems. Over a two-day period, participants must solve six problems in nine hours.

For Further Information: Write to Walter Mientka, Executive Director, American Mathematics Competitions, Department of Mathematics and Statistics, University of Nebraska, Lincoln, NE 68588-0322. Telephone (402) 472-2257.

PART III:

Advice From Hindsight

SURVEY OF NOBEL LAUREATES AND MEDAL OF SCIENCE WINNERS*

```
164    surveys sent
 87    returned
+  2   interviewed
 89                                    54 % return
+ 12   (deceased or unavailable)
101                                    60 % return†
```

Question 1—Did the mature scientists join and/or lead science clubs?

```
 39    — No
+ 27   — No, didn't exist
 66                                    75 %

  1    — Yes
+ 20   — Yes, with explanation
 21                                    24 %

+  2   — No answer                     2 %
 89
```

Question 2 —Did they enter and/or win science fairs?

```
 51    — No
+ 26   — No, didn't exist
 77                                    87 %

  8    — Yes, with explanation         9 %

  3    — No answer                     3 %
+  1   — Don't know                    1 %
 89
```

Question 3—Did they enter and/or win science competitions?

```
 44    — No
+ 25   — No, didn't exist
 69                                    78 %

 15    — Yes, with explanation         17 %
+  5   — No answer                     6 %
 89
```

Question 4—Value of competitions and influence on career

```
 23    — Undecided                     26 %
  8    — Undecided (irrelevant)         9 %
+ 13   — No answer                     15 %
 44                                    50 %

  8    — Good
+ 24   — Good, with reservations
 32                                    35 %

  8    — Bad
+  5   — Bad, with reservations
 13                                    15 %
 89
```

*With one exception (noted †), the percentages are based on the 89 responses.
Rounded percentages may not total 100.

EIGHTY-NINE NOBEL LAUREATES AND MEDAL OF SCIENCE WINNERS COMMENT ON CLUBS AND CONTESTS

Deborah C. Fort

As NSTA staff began collecting the information about the science and math competitions and other activities that make up the bulk of this volume, a number of persistent questions emerged. The most disturbing among them was this one: Are competitions—even well-run ones—sound educational practice? Second, does participation in competitions have any measurable effect—negative *or* positive—on future career choice?

In concrete terms, did the youth who joined the physics club, or who won the science fair, or whose project placed in the National Junior Horticulture Association competition, or who was a Westinghouse Science Search finalist, or who led the U.S. Mathematics Olympiad team to victory in Europe, become in adulthood a scientist or mathematician? Or were other factors as important, more important? Are science and math competitions, in general, as fundamentally irrelevant to students' eventual professional life as are, for instance, school athletic programs or language clubs? (It's rare that the high school quarterback becomes a professional football player, that the president of French Club goes into linguistics.) Is it possible that competitions are not merely irrelevant but even inimical to a future in science or math?

These questions call for a larger study than NSTA resources could encompass at this writing. But we did think of a place to start and, accordingly, in mid-October, 1989, NSTA sent out 164 surveys to a list of men and women intended to comprise all living U.S. Nobelists and Medal of Science winners. By late December, we had returns from over half those solicited: 87 responses, including the 20-page biography one Nobelist chose to submit, (89, including the two more from Laureates who agreed to serve as guinea pigs with telephoned replies to our four questions); plus another 12 surveys returned because the intended recipient was recently deceased or otherwise unavailable. The results (summary follows) were thoughtful, instructive, often quirky, and always interesting. In briefest form, while a large majority of respondents had not joined or competed, only about 15 percent thought such activities were detrimental, 35 percent believed them generally good, and about half were undecided about their value. NSTA is most grateful to those who took time from full and busy lives to offer detailed, revealing answers.

In statistical form the results follow:

RESPONSES RECEIVED FROM 89 U.S. NOBELISTS AND MEDAL OF SCIENCE WINNERS*

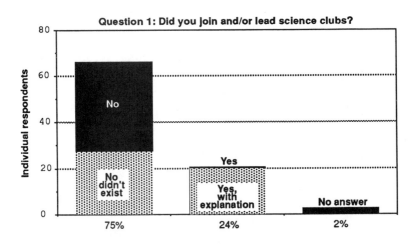

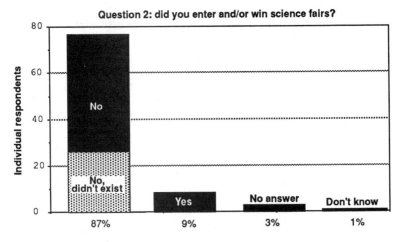

164 surveys mailed fall 1989. Rounded percentages may not total 100.

About three-quarters of the award-winning scientists who returned surveys had neither joined nor led science clubs in their youth, 27 of them because such clubs—along with science fairs and larger competitions—simply did not exist when they were in high school. In the United States, Science Service estimates that science clubs and fairs originated well before the Second World War.

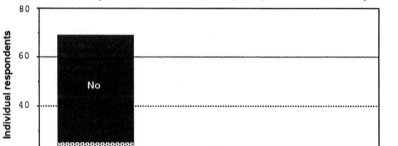

Question 3: Did you enter and/or win national, state, or local science competitions?

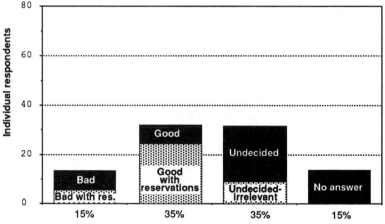

Question 4: Are science competitions valuable and what influence did they have on your career?

Responses to other surveys NSTA sent as part of this project reveal that, in 1935, with the cooperation of the Vegetable Growers Association of America, a precursor of the national science competitions was organized in the form of a vegetable growing, judging, and identification contest. In 1939, the National Junior Horticulture Association took over this contest, which has grown and diversified into the events described above in this book (page 110). The earliest of the nationwide science contests, the Westinghouse Science Search, began in 1942; the oldest of the international competitions, the International Science and Engineering Fair, opened for the first time in 1950.

Given the fact that the 85 respondents were born between 1888 and 1940 and were mostly well out of high school by the 1940s, the large percentage who didn't participate in clubs, fairs, or other competitions is not surprising. And, given the scientist's preference, even need, for evidence, the nearly half of the sample who had no clear opinion about the value of science and math competitions is also easy to understand. There are, however, other ways to gain experience about such contests besides competing in them.

For example, the boyhood of Glenn T. Seaborg (Nobelist in chemistry) in Michigan's Upper Peninsula was marked by an unfamiliarity not only with science contests but even with radios. Though Seaborg first "turned on to science" in his junior year through the work of an inspired high school teacher, Dwight Logan Reid, he reached an informed opinion on the value of competitions through other means. Seaborg's nearly 30 years of service judging the Westinghouse Science Search leaves him, like 35 percent of his prize-winning colleagues, with a fundamentally positive view of science and math contests: He believes contests to be "exciting, constructive, and stimulating" and the source of "friendships that last."

The (admittedly) "limited" experience of geneticist Joshua Lederberg, also of the wrong generation to have participated in science fairs or other contests, led him to another conclusion, along with the 15 percent of respondents who find science fairs to be either entirely or in certain major aspects injurious. Lederberg has had noncombatant contact with science competitions, both as a parent and as a judge—the latter role, he says, he has long refused to play because organizers refused to let him be a "maverick." Lederberg is left, overall, with a negative impression. He summarizes:

I have a hunch—only a hunch, not based on evidence—
that the runners-up would do better in future years than
the winners because the criteria, the qualities on which
kids are judged, often go against real creativity. [There's]
the problem of overreach, of overly ambitious, elaborate,
glossy projects winning whereas there can be real creativity
in failures.

On the other hand, as Seaborg acknowledges that badly run
competitions could be damaging, Lederberg points out that while
some real creativity (but also some real incompetence) can exist in
failed projects, "some of the winners are marvelous kids."*

SCIENCE INTEREST GROUPS OR CLUBS

"But science *clubs*?" asks Lederberg, a graduate of Stuyvesant High
School in New York City. "They are a totally different matter. That's
peer, not outsider, judgment." Science clubs were very much around
when Lederberg was in high school, and he hastened to become
active in four of them, joining the biology, chemistry, math, and
premed clubs.† Asked if he ever slept, Lederberg laughed, "Not
when I was a kid."

About a quarter of Lederberg's colleagues were also active in
science or math clubs—nearly twice the number who participated in
fairs or other competitions. Three joined math clubs; Herbert A.
Hauptman, 1985 Laureate in physics, elaborates that the group he
joined at 15 was actually a "problem-solving" club:

Problems in math (mostly geometry) were suggested by a
faculty advisor, and we student members attempted
solutions and reported our progress...it was a lot of fun...I
believe it played no role in my future career; I was already
in love with math and had been for as long as I can
remember.

Nine other future scientists were members of general science
clubs during their elementary and/or high school years, three
serving as officers. Others joined more specialized groups: Herbert
Gutowsky, for example, served as secretary of his school biology
club. Two, growing up in less isolated environments than Seaborg's

*Lederberg would like to see a study done on the number of runners-up (compared
to winners) in national contests who become scientists or mathematicians.

†Because like Lederberg, a number of the Nobelists and Medal of Science winners
joined more than one group, the number of clubs described here slightly exceeds the
numbers of joiners tallied graphically above.

Ishpeming, Michigan, joined amateur radio clubs; one, a camera club; another, a biology club, where he performed experiments of his own choosing; two, high school astronomy clubs, where one ground a mirror and lens and constructed a telescope. Some of the groups were either more humble or more exotic. Sheldon Glashow, who won his Nobel a decade ago, was a member of the "protozoology squad," a service organization whose members cleaned the biology lab, and a science fiction club in which he served as features editor of its off-campus fanzine, *Etaoin Shrdlu*.*

Yuan Tseh Lee, 1986 Nobel prize winner in chemistry, found out at 16 what many teachers discover only when they enter their own classrooms after college—that is, the best way to learn something is to teach it. Lee writes:

> In ninth grade, I headed a "review club" of physics and chemistry to help my classmates review...for entrance examinations [for] senior high school. I learned so much during the year, and I was told that I [understood] most of the physics and chemistry that were offered in senior high school.

One Nobelist, like Lederberg, joined a number of clubs—chemistry, physics, and math—where he heard lectures, worked in laboratories, and attended social meetings. Another, Philip Anderson—who, again like Lederberg, holds both the Medal of Science and the Nobel Prize—joined none; he says that while such clubs "hardly existed" in his boyhood, he would not have participated anyhow: "We placed great value on independence and *non-joining*, and such things would have been seen as demeaning."

SCIENCE FAIRS

Even fewer of our respondents—only about one in ten—entered science fairs in their youth than joined clubs. Twenty-six award winners couldn't enter fairs for the same reason that they eschewed clubs and contests—they were born too soon. None of the 51 others who didn't participate shared their reasons, many reserving comments for the last open-ended question on the survey (see below). The eight who did enter fairs provided explanations. One,

---

*The name of the magazine reproduces the two sequences of letters on the two left-hand vertical rows of a linotype keyboard. When a careless or unlucky linotyper—in tandem with a similar proofreader—accidentally runs a finger down these rows, *etaoin shrdlu* appears in print. See Glashow's completed survey reproduced in the appendix below.

who later eloquently argued against "competitive drive," noted that the science fair in which he participated "was not a *competition*"; nonetheless, in it he received a "first citation."

Whatever that citation signified, this future Nobelist was joined by only one of the other respondents with a first-place science fair award—Glenn W. Burton, now a research geneticist with the U.S. Department of Agriculture, whose collection of native grasses took first place in a county fair. Three other fair contestants have these results to report: Yoichiro Nambu "thinks" he received a prize for the geographical model he submitted at age 10; Jack Steinberger, motivated by a "very nice chemistry teacher," entered an exhibit of different chemicals but can't remember how his submission was rated; however, Glashow recalls his third-place prize in elementary school for the stroboscope he made and demonstrated. George D. Snell won no science fairs; however, he took first place in a *math* competition and won a prize in *art*.

NATIONAL, STATE, AND LOCAL COMPETITIONS

Glashow, one among five Westinghouse Science Search Finalists* who went on to win the Nobel, obviously numbers among the 17 percent of Laureates and Medal of Science respondents who entered contests organized above the school level. Like the other 15 respondents who report having participated in the national-level competitions, Glashow elaborated on the nature of his project, the substitution of selenium for sulfur in hydroponic plant growth, but somewhat uncharacteristically reports that he "received *no* assistance, lab space, teacher time, etc." from his school. A number of other scientists like Gutowsky, Seaborg, and Steinberger specifically noted encouragement from high school teachers.

Of the now familiar 25–26 prize winners whose generation largely excluded them from such contests, two wished they could have entered. Verner E. Suomi writes that he had "no opportunity but it would have been fun, and [a competition project] helps focus on a thorough, complete effort." Burton wrote that in his four-teacher high school, there were no contests offered but that he "would have entered if he could." A third laureate reported two awards based on state Regents' grades, not, he said, true competitions. Sixty-nine others, about the same number who didn't join clubs (66) but 11 fewer than those who didn't participate in

*Other Laureates and Medal of Science winners *entered* the Search but were not among the 40 *finalists*.

fairs (77), simply noted that they did not compete.

Six respondents besides Glashow participated in the Westinghouse Science Search. Like Glashow, Roald Hoffmann was a finalist, an event that—in spite of his "inherent doubts about competitions"*—was enormously influential in shaping his professional life. Hoffmann writes, "For me, success in the [Science Search] led to a summer research job and to a career in chemistry." David Baltimore "thinks" he received an honorable mention in the Search in which Baruch S. Blumberg recalls that he won "some prize." Blumberg also placed in the General Electric Science Competition. Medal of Science winner Donald E. Knuth remembers achieving honorable mention in the Search for a mathematics paper that led to a publication in graduate school; he also won a prize ("most original presentation") from a Wisconsin Junior Academy that led to a paper in *Mad Magazine*. As an eighth grader, Knuth, a dedicated participant, also won a contest to make the most English words from the letters of "Ziegler's Giant Bar." Knuth found over 4,500 words, more than twice the number on the judge's list. Two other respondents, however, report participation but no award in the Search.

Two respondents—Ralph B. Peck and Neal E. Miller—took first place in state chemistry competitions; another came in third in his high school science contest. Two others—Yuan Tseh Lee and Mildred Cohn—participated in high school math competitions. Cohn writes,

> I was a member of the math team (5 members in a school of 5,000 students)...We competed with other high schools of New York City. Our team did not win the city-wide competition, and I do not remember where we placed.
>
> Since I did not compete, except in mathematics, I don't think it had any effect on my career. As a footnote, I should add that I was the only girl on my high school math team and that probably reinforced my idea that I could hold my own in a man's world.

Finally, Nambu, at the age of about ten, took part in an abacus competition, while George E. Mueller, now president of the International Academy of Astronautics, was active in model airplane competitions, "usually finishing third or fourth."

---

*See below for Hoffmann's further discussion of the pros and cons of contests.

SCIENCE COMPETITIONS: GOOD, BAD, OR QUESTIONABLE?

IN REVERSE ORDER: QUESTIONABLE. About half of the respondents were undecided: choosing to leave the question unanswered (15 percent), citing insufficient data either from personal experience or from research about competitions to make an informed judgment (26 percent), or acknowledging so little personal impact as to make the query irrelevant (9 percent).

Individuals in the last two categories often offered thoughtful reasons for their inability to make decisions about the value of entering competitions. Finding competitions irrelevant, for example, a number of Laureates and Medal of Science winners tried to explain what they thought *did* impel them toward science. Some cited inspired teachers and experiences in schooling. Howard M. Temin's summer at the Jackson Memorial Laboratory in Bar Harbor, Maine, for example, was essential, as was then high-schooler Robert N. Noyce's opportunity to take physics at Grinnell College in Iowa. Others remember early interest in nature, among them James V. Neel (who had a butterfly collection), James D. Watson, an ornithologist, and Ernst Mayr ("I was an ardent young naturalist"). With "another kid," 12-year-old George Wald

> had become absorbed in very simple and crude—hence inexpensive—electrical experiments. [They were] all done at home. We devoured each issue of Hugo Gernsbach's magazine, *The Electrical Experimenter* (later *Science and Invention*). I built a radio receiving set in my bedroom, with a galena crystal and phosphor bronze "cat whisker," made a tuner out of an empty oatmeal carton wound with number 24 magnet wire, and strung an aerial up on the roof. All the materials were bought with what money I could scrounge—no school connections, whatever. My prize: All the kids on the block piled into my bedroom for the World Series!

A number of respondents note the influence of their educational environments, some citing particularly their parents' encouragement and guidance. Many had to wait until well into or even beyond their secondary years to uncover an interest in science.

Frederick Seitz, among those scientists who found the effect of competitions overall unresolved, even if not vital to the course of their particular careers, expressed sentiments that came up frequently among the respondents. Seitz writes, first,

...beginning in grade school I had an absorbing interest in science and knew that if possible I would select a scientific career. As a result, I read avidly almost anything that came my way, often reaching for things that were somewhat above my head.

[and then]

I do not believe that science competitions are harmful but also do not feel that they are essential as a stepping stone for those who will follow scientific careers. In my own case, the competitive side of science was, initially at least, secondary to the lure of the subject itself.

Recent Nobel prize winner Jerome Karle, the final spokesperson for the nearly half of NSTA's respondents who have basically came down as "undecided" about the effect of competitions upon fostering pleasure and/or talents in science, is typical of the group as a whole in a number of ways. He is thoughtful, evenhanded, serious, generous, providing a careful single-spaced page of prose to consider the question. He offers this extended response, although "evidently competitions played no role in my development as a scientist." Following are excerpts:

The virtues of science competitions are worth considering. They obviously stimulate participation in and concentration on some technical subject. They help some students in rural areas make contact with the outside world and perhaps gain some confidence in their relative abilities. It is occasionally possible for a selected group to acquire funds at science fairs for further education. On the down side, parents, friends, and friends of parents may at times afford unfair competitive advantages and even contribute so much that the student may not comprehend what is going on. Another question concerns whether losers in competitions are stimulated to continue or are discouraged. If the latter is so, it may not be worth encouraging a handful at the expense of the many who might otherwise have developed their talents and interest. Science competitions do not identify all individuals who have a significant future potential in technical subjects. There may be more effective ways to spend the time and effort.

In the context of development as a scientist, I greatly benefited from the high quality, well-disciplined educational environment in New York City as I was growing up...It was an environment in which scholarship was very highly regarded and the building of character, proper behavior, and respect for individuals were inherent parts of the educational system...It was a time when children were not deflected from a proper course by too many economic advantages, too many distractions, and much too much media irresponsibility in presenting low standards of speech and personal behavior as though they were perfectly natural and perfectly all right...

The aspects of a good education that were of prime significance in my development, however, are, except for isolated instances, very badly needed by our present society. It is they that help create decent people with proper priorities and standards. It is they that provide the stimulus for encouraging intellectuals of all callings, including technical people.

COMPETITIONS: DOWN WITH THEM. Those 15 percent of replying Nobelists and Medal of Science winners who—after sorting through the data and forming conclusions—oppose contests tend to express themselves both more firmly and more concisely than the larger undecided group (almost half of the respondents).

Eight found competitions virtually without redeeming value. They are "too goal oriented" and "thwart independent and original research," writes Barbara McClintock. Concurs Hauptman, "I do *not* endorse the idea of holding competitions. I do not believe they are a healthy stimulus to motivation." "I am skeptical about their value," writes Linus Pauling. Adds Wald,

> I do not endorse science competitions. The entire tenor of our society is over-competitive. That surely contributes to the amount of cheating now going on in science. The only advantage I see to having any of this is that it may make available time, space, and simple equipment for a few kids who are eager to experiment. They don't need contests and prizes to urge them on. In fact, there is not much you can do for them but good teaching and seeing that they get as good and as much education as they can absorb.

Two other critics refer to Watson to attack science competitions, though in different contexts. Comments one, competitions tend to "attract students who do not have the requisite originality and independence of mind. How many *great* scientists started this way?" he wonders. "Do you see Jim Watson of *The Double Helix* doing a science fair?" Writes another,

> I am not persuaded that science competitions serve the cause of science. Despite the impression that Watson leaves

in *The Double Helix*, competition is not a mainspring of science, though it is of course a factor. I do not believe that scientists do their best work in the spirit of competition and to emphasize this element early on may well be inimical to the health of science.

But Watson himself offers another perspective. "Science fairs," he explains, "were not held in the middle 1940s." He adds, however, that "the main thing is an opportunity to do science while still young."

Four other respondents joined Lederberg in seeing science competitions as being fundamentally less helpful to students than misleading about the nature of science. Worried about the perhaps "misplaced" emphasis on competition, one respondent wonders whether competitions might be less "effective [than] providing funds for research and science activities." Lederberg would like competitions held in a "realistic framework, resembling in some ways that in which scientists work" and hopes that "kids could get a chance to have contact with scientists, with mentors, in a realistic time frame." Similarly, Frederick C. Robbins would "encourage curiosity and exploration" rather than competition.

COMPETITIONS: ON THE UP SIDE. The remaining 35 percent of the respondents found science competitions either entirely or mostly a positive way of involving students in science. Of these 32 respondents, 24 retained minor reservations about the value of competitions, while 8 found them useful and productive almost without caveat.

A number of the latter group drew certain analogies with athletics—both positive and negative. Edward O. Wilson was not alone in being glad that there were some competitions available in which "(unlike athletics) I could excel." Miller noted a difference between team sports and science competitions that he would like to narrow:

> Competitions should be arranged so that winning brings credit to the school and/or the team, as is the case with athletics. The outstanding student often is unpopular because by comparison he or she makes classmates look bad; the outstanding athlete helps the school to win.

The youngest of the respondents, Kenneth G. Wilson, would also like to see experimentation with format, "with special attention to issues such as team versus individual competitions and [to] the question of how broad the reach should be both in terms of numbers of participants (enough to include all future scientists?) and numbers of winners." The suggestions of all three of these last respondents make note of the collaborative nature of science so well explored by Derek de Solla Price in *Science Since Babylon* (1961/ 1975) and *Big Science, Little Science and Beyond* (1961) and by Harriet Zuckerman in *Scientific Elite: Nobel Laureates in the United States* (1978). Perhaps the team aspect of some of the olympiads is a step in this direction; however, while cooperation is important in science, there are also clear leaders whose genius is essential to discoveries, formulations, or other breakthroughs. One does not give equal credit to Jonas Salk and his lab assistant for the vaccine! The issue of true *teamwork* versus independent efforts that turn out to be complementary, however, is much too complex to be more than merely raised here.

Several respondents expressed concern about the feelings of participants who do not come in first. For example, Mueller writes that, while competitions are "very worthwhile, one needs to be careful to adopt a format which does not turn the losers off science but rather encourages every participant—win, lose, or draw—to continue the pursuit of a scientific education." On a similar note, Dudley Herschbach finds fairs and contests "probably good, because they encourage students to undertake projects of their own." Competitions also, he continues, "may increase awareness of parents and [the] public. [Organizers] should be generous with prizes, honorable mentions, etc., to avoid disheartening the majority of the entrants."

Hoffmann summarizes both the strengths and potential weaknesses of such competitions. In spite of his "inherent doubts" mentioned above, he concludes that

> Anything that encourages youth participation in science is to be encouraged...I worry that [competitions] are more the outcome of teacher interest than student interest; I worry that competition is less consistent with science than cooperation. I especially worry whether they serve to *identify* science talent...

But, even with these hedges, Hoffmann believes that his success in the Science Talent Search started him toward his career in chemistry. Approaching the question of the value of competitions from a slightly different angle, Roger C. Guillemin worries less that those with talent enter because "those who participate are *already* interested in science. The problem is to motivate those who are not so interested..."

Those 32 respondents who found little to criticize in science competitions and other organizations also found a great deal to praise. "They seem a good way to involve the already committed and maybe to excite the uncertain," writes Baltimore, sentiments echoed by a number of his colleagues, many of whom also believe that competitions raise public awareness of science. Knuth is even more enthusiastic:

> I believe all such outlets encourage creativity. Clubs may work best only if "attractive" people are members. I personally think it is better to have enriching activities like this instead of advancing students to higher grade levels. The same goes for all subjects, not just science ([take, for example,] spelling bees). In seventh grade a bunch of us spontaneously challenged each other at trying to diagram...sentences grammatically; that experience was later to help me in computer programming.

Several respondents approved the unusual kind of learning that can take place as students prepare projects. Writes William H. Pickering, "In most cases, the students learn much more about their topic than would have been the case had they not entered...." Science competitions, according to Daniel I. Arnon, can "direct the thoughts and focus the intellectual and mechanical energies of the participants on a specific project to illustrate a scientific or engineering principle with clarity, ingenuity, and originality...." Lewis H. Sarett spoke for a number of respondents when he speculated that competitions "should help motivate a fraction of scientists-in-the-making." Arno A. Penzias found entering contests "a nice break from the creation of expected answers [for] an exam that seems to drive many of our bright students, unfortunately." Similarly, Suomi writes, "I think science competitions are valuable because they are opportunities in informal education. In my instance, it was the informal education that made me strive for the formal education."

One of the Laureates with personal experience with both clubs and competitions summarized his experience as follows: "I think [competitions] helped me. [They] stimulated my interest in science and taught me an important lesson about experimental science—how to collect and analyze and report the data. The clubs and competitions improved my interaction with other students who had similar interests." The Nobelist added, however, two statements that were frequently echoed by other respondents, one positive, one negative. His science activities were also shared and supported by his teacher "who was very important." But—like many of his colleagues—"I oppose the overparticipation of parents and other adults."

In another frequently mentioned note of caution, Lyman Spitzer, Jr., points out that evaluating the effects of contests is still largely in the area of hypothesis and guesswork—the hard data are yet to be collected. "I have always assumed that science competitions served a useful purpose," he writes, "in building up interest...among the general public and...in encouraging young people to enter science as a career. However, I have no detailed evidence to support this very plausible assumption." Westinghouse finalist Glashow, however, offers a convincing sample of one: "Such competitions," he writes, "represented the only encouragement I ever received from the ed-biz establishment to pursue science as a career." Science Service statistics show that 95 percent of former Search winners choose some branch of science as their major field of study, more than 70 percent earning PhDs or MDs.

So, while nearly half of the 89 respondents (44) came down as undecided about the value of competing in science and mathematics, more than twice as many of those taking a stand were in favor of contests (32) than against them (13). The ambivalence expressed here may be emblematic of something fundamental in many successful endeavors in science and mathematics: Breakthroughs are both the fruit of genius, which is profoundly individual if not specifically competitive, and of cooperation among helpers. The duty of the scientist is similarly twofold: First, to discover, to uncover, new knowledge—often a solitary task but often as part of a team, and then, to share it, to collaborate with peers, through publication in journals or orally at professional meetings.

REFERENCES

De Solla Price, Derek. (1961/1975). *Science since Babylon* (enlarged ed.). New Haven: Yale University Press.
De Solla Price, Derek. (1961). *Big science, little science and beyond* (rev. ed.). New York: Columbia University Press.
Zuckerman, Harriet. (1978). *Scientific elite: Nobel laureates in the United States*. New York: Free Press.

Deborah C. Fort, a Washington-based free-lance writer, editor, and teacher, has most recently served as NSTA association editor for Gifted Young in Science: Potential Through Performance *(coeditors Paul F. Brandwein and A. Harry Passow, contributing editor Gerald Skoog).*

PART IV:

Appendices

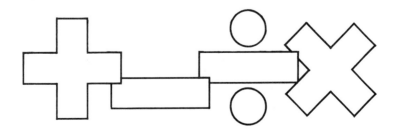

QUESTIONNAIRE

Returned by Science Club Advisor
Susanna Chinouth

Science Clubs Survey

My ___middle___ school, __Liberty Bell Middle School__*
(level—elem., m/jh, secondary)   (name of school)

__Liberty Bell Blvd.__

__Johnson City__, __TN__ __37604__ __461-1621__
address—city   state   zip code   office telephone

with __35⁺__ students has had a science club for __this__ years.
(number) __and parents__   (number)

In 1988–1989, the club had _____ members and met_____times.
          (number)   (number)

It_____ a formal constitution, which _____ enclosed.
(has/has not)   (is/is not)

The club's faculty advisor, __Susanna Chinouth__ spent __20__ hours
         (name)   (number) __so far__

advising, guaranteeing safety, other __Planning organization__
__protocol, arranging guest speakers,__
__Communicating with adult officers__ during this year.
   (please specify)
__S.P.A.S.E. (Students & parents for the advancement__
The club's primary activities were __of science education)__
1. __Wildlife Resource speaker on "Mammals of__
                __Southern Appalachians"__
2. __Star-gazing party with University professor__
3. __Money-making projects- for science programs__
4. __We have applied for 5 VISITING TENNESSEE__
   __SCIENTISTS__
5. __HELP WITH LOCAL SCIENCE FAIR__
Other information__This is a Booster club meant to raise__
__money for the Science department and__
__to enrich students on varied science topics.__
__We are expecting 70⁺ people at our next meeting__
*Please return to NSTA, 1742 Connecticut Ave. NW, Washington, DC
20009

QUESTIONNAIRE
Returned by Science Club Advisor
Pat Foy

Science Clubs Survey

My _Secondary_ school, _Charter Oak High_ .
(level—elem., m/jh, secondary)   (name of school)

P.O. Box 9

Covina CA 91723 (818) 915-5841
address—city   state   zip code   office telephone

with _____ students has had a science club for _5_ years.
      (number)   (number)

In 1988–1989, the club had _25_ members, and met _16-20_ times.
                           (number)   (number)

It _does not have_ a formal constitution, which _is not_ enclosed.
   (has/has not)   (is/is not)

The club's faculty advisor, _Pat Foy_ , spent _100_ hours
                              (name)   (number)

advising, guaranteeing safety, other _prepping for science_
competion, contacting/reservations for
activities _____ during the year.
        (please specify)

The club's primary activities were _envolvement_
to increase student activities in science
related activities : Tidepooling, whalewatching
visit to Natural History museum, Catalina Island
Marine Institute, quest speaker : Science & college
major. Competition in L.A. Co. Environ.Ed Fair

Other information
The club is an informal group w/a varying
about of students. Anyone on campus was able to
attend activities - No membership Requirement
(i.e. Enrolled in a science class) is involved.
*Please return to NSTA, 1742 Connecticut Ave. NW, Washington, DC 20009

QUESTIONNAIRE

Returned by Competition Sponsor
Duracell NSTA

COMPETITION SPONSOR SURVEY

Please help the National Science Teachers Association by responding as fully as possible. If you have a brochure or entry form that answers any or all of these questions, please send it along with the questionnaire. Write as much as you like, using the back of the survey or separate sheets, if necessary.

1. What is the official name of the competition?
   Duracell NSTA Scholarship Competition

2. What is the history of the competition? Who started it and in what year?
   Sponsored by Duracell USA and administered by NSTA, the competition is in its 8th year.

3. Who funds the competition? Please list all sponsors.
   Duracell USA

4. Are there any now-famous past winners?
   Not that we know of.

5. Who may enter?
   Students in grades 9-12 residing in the U.S.

6. What is the nature of the projects? Essays? Research? Inventions? Tests? Displays? Other?
   Invent and build a battery-powered device. Submit a 2-page essay describing the device, a wiring diagram, a photo and an official entry blank.

7. Are there other criteria for entrants, such as minimum GPA or science GPA?
   No

8. What are your deadlines for application and entry?

   Vary yearly. Usually in late January, early February.

9. What are the awards and prizes?

   One $10,000 scholarship, five $3,000 scholarships, ten $500
   scholarships, and 25 cash awards of $100. Teachers of winners
   receive computers and NSTA publications.

10. Will the entrants have to travel? Expense paid?

   Six top winners receive expense-paid trip to NSTA National
   Convention in the spring for the Awards Ceremony.

11. Please provide the names and addresses of five winners and/or
finalists and their teachers and five competition judges.

   We are conducting a survey of our own this year. Please call
   me and I'll tell you about our survey. Maybe we could
   incorporate your questions and share the results with you.

QUESTIONNAIRE
Returned by International Physics Olympiad Contestant
Steven S. Gubser

For its upcoming guide to math and science competitions and other opportunities, the National Science Teachers Association needs some details about how you did your project, from start to finish. Can you share any hints that helped you or pass along any suggestions to other students and teachers? Write as much as you like, using the back of the survey or separate sheets, if necessary.

1. What is the name of the competition you entered?
The Twentieth International Physics Olympiad

2. Please describe your project briefly.
I was one of five members of the U.S. team that went to Warsaw, Poland. Thirty-three countries attended the competition, bringing about 150 students. All 150 took a five-hour theoretical exam and a five-hour lab exam--both written. My combined score on these two exams was the top score in all the world.

3. Where did you get the idea for your project and how did you develop that idea?
Another student in my high school went to the same competition four years before I did. I talked to the physics teacher he had had and decided to enter the competition myself. After three successive tests and a week of training at the University of Maryland, I qualified for the competition.

4. How much time did you spend on your project? How many hours per week and how many days, weeks, or months?
I spent ten hours a week during late April and May. Then in June and early July, I spent about 40 hours a week.

5. How much help did you get from teachers, mentors, parents, or others? (Please provide names if possible.) How did they help--with suggestions or ideas for your project, moral support, or other guidance?
Arthur Eisenkraft and Larry Kirkpatrick--the academic directors of the U.S. Physics Olympiad--gave me extensive help during the training camp at the University of Maryland. I also got help from my high school physics teachers Neil Mackie and Pat Ryan. At the University of Colorado, Professor Al Bartlett and Professor John Taylor gave me a hand, too.

6. Where did you work on your project? What special facilities did you need, if any?
I went to a training camp at the University of Maryland and used their lab facilities. I also did labs at the University of Colorado with the help of Merle Ware and Jerry Leigh. The rest of the work I did at home.

7. How did you hear about the competition?
See #3.

Science and Math Events: Connecting and Competing

8. How did you feel when your project was done and then when the competition was over?

Great. I relished the opportunity to increase my knowledge of physics and the experience of meetin physics students of other nationalities.

9. Why would you or wouldn't you recommend this competition to other students?

I would recommend it to any student interested in physics because it covers and surpasses the whole curriculum of college freshman physics.

10. At this time, four years/months after the contest, do
   (number) (circle one)

think you will eventually end up working as a mathematician or scientist? If so, do you have any idea of what specialized field you might enter?

I think I will become a theoretical physicist. I do not have a specialty picked out, but I am presently very interested in relativity.

11. Any other observations?

12.   Student's Permission

You may/may not publish my responses.
   circle one

You may/may not attribute them to me by name.
   circle one

Name Steven S. Gubser   Signature _Steven S. Gubser_
      please print

Telephone number (303) 741-0429
             (area code)

13.   Parent/Guardian's Permission

You may/may not publish my son/daughter's responses.
   circle one

You may/may not attribute them to him/her by name.
   circle one

Name Nicholas J. Gubser   Signature _Nicholas J. Gubser_
      please print

Telephone number (303) 741-0429
             (area code)

QUESTIONNAIRE

Returned by Science Teacher
Carly Erikson

The National Science Teachers Association is looking for information to provide teachers about whether to encourage their students to enter math and science competitions. We received your name from sponsors of a competition in which one of your students excelled. Please help us advise other teachers by responding as fully as possible. Write as much as you like, using the back of the survey or separate sheets, if necessary.

1. What competitions do you encourage students to enter? Have any of your students won any of them? Which one? AAPT? Metrologic Physics Contest. We attend the Colorado Physics Bowl every year and enter both the team and individual competitions. I encourage my students to try NASA's Name the Shuttle Contest, Duracell, and Westinghouse competitions though screen the final entry. I entered anything else costing a minimum dollar amount.

2. How do you hear about the competitions and come to be involved in them? I am a member of NSTA and AAPT and read all literature supplied... especially the trade journal The Physics Teacher. I also attend national conferences and the Woodrow Wilson Institute Super Physics 1988.

3. Is there a special program at your school that helps you and your students with competitions? How does it operate?

Hewlett Packard coordinates with us in a Visiting Scientist program. My visiting scientist and I have worked together for four years. He has a BS in mechanical engineering, an MS in electrical engineering and loves teenagers!

4. How do you encourage your students to get involved and to keep going once they are working on their projects?

I try to steer them towards projects they have the ability to do. (Very few of my students can do a decent Duracell project.) This cuts down on frustration.

5. How much do you help students? (Specific examples would be illustrative.) Did you help them decide on a project, guide their progress, help them with research, writing? In what ways?

I help them very little. We do train teams for the Physics Bowl (sample questions and competitions). For others, I suggest readings or problems with subjects they are weak in. We de-emphasize teacher involvement (which is why I'll probably never have a student win Duracell of Westinghouse!)

6. How much, if any, time do you schedule for helping students work on competitions? How many hours per week and number of weeks, months?

We train for Physics Bowl two times per week for one hour, for two months.

7. Who else helped the students? Other teachers, parents, business people, or scientists from outside the school community? How did they help?

See description of Visiting Scientist on question #3.

8. How did your students react to contests overall? How did you handle discouragement? Was there one student who was especially challenging or rewarding?

We enjoy contests. Discouragement is handled analytically by looking at strengths and weaknesses. We also celebrate the end with lunch or pizza party.

9. Did you face any problems relating to your students' physical safety as they worked on their projects? If so, how did you handle those problems?

No. Safety is a prime issue; I avoid any risks.

10. Would you recommend to other teachers that their students enter competitions? Why or why not? Please be specific.

I believe it depends on the teachers attitude. First, the kids compete (teachers don't--we coach), so I believe teacher involvement should be minimal. Also competitions should be a learning experience and a chance to be with like kind.

11. Do you have any other suggestions or observations to offer?

I wish there was a way to a) control the politics of national competitions. There seems to be a bias against we western states. b) decrease teacher involvement c) find out the results. For example, what did the winners of the NASA contest do that we did not?

QUESTIONNAIRE

Returned by Westinghouse Judge
Nobelist Glenn T. Seaborg

For its upcoming guide to math and science competitions and other opportunities, the National Science Teachers Association needs some details about how you judged a competition. Can you share any hints that helped you or pass along any suggestions to other contest judges? Rather than specific rules and criteria as stated in official competition regulations, we are looking for insight into how you made your judging decisions--what made one entry better than another in your mind. Write as much as you like, using the back of the survey or separate sheets, if necessary.

1. What is/are the name(s) of the competition(s) you have judged?

Science Talent Search (1964- 1989)

2. What special qualities did you look for in a winner? Did you consider student effort, or was the quality of the finished project your main criterion?

Breadth of knowledge and understanding of science as determined by an oral interviewer.

3. What is the range of entries you have seen, from most to least complex or sophisticated?

Broad range of competence as determined by interviews--from very little to very sophisticated.

4. Did the competition you judged reject some projects outright? If so, what percent of the entries were so dismissed? Why?

No

5. How were the projects presented? Were the students present? Did they explain orally what they were trying to achieve? Or was their anonymity protected? Please tell us what you think is the best method of student presentation of science and math competition entries.

Students were present and were interviewed individually, with their identity known. Best method is use of interviews.

6. Would you encourage teachers to get their students involved in competitions? Why or why not? If you would, how would you do so?

Yes, hands-on experience is valuable. Send information on competition to teachers.

7. Have you seen evidence that students who don't win feel unhappy? If so, is this inevitable? Even desirable ("part of maturing")? Would you like to offer any suggestions to help soften the experience of failing?

Most students seem to understand that the winners were better prepared and deserving.

8. Can you give any special insight into how you judge science and math competitions?

I look for basic understanding and originality.

Please return to NSTA, 1742 Connecticut Ave., N. W.,
Washington, DC 20009 Attn: Deborah C. Fort
Deadline: November, 1989

QUESTIONNAIRE
Returned by Nobelist
Sheldon Glashow

Name Sheldon Glashow   Address Lyman Lab of Physics, Harvard Univ.

City Cambridge   State and ZIP MA 02138
MELLON PROF. of the SCIENCES
Title_____Affiliation Physics-Harvard

Award and Date conferred Nobel-1979_____Date of Birth *DEC 5 1932*
(Nobel and/or Medal of Science)

Please verify the above information and fill in the title by which you
like to be known.

(In each case below, insofar as memory permits, please indicate whether
you participated in any of the following activities during your school
years.)

In (elementary and/or secondary) school, I was a member and/or officer
of science club(s) _____*YES*_____. If yes, please briefly describe
                        yes/no
its/their activities, specifying your role and approximately how old
you were at the time.

H.S. "PROTOZOOLOGY SQUAD" A SERVICE ORGANIZATION
to CLEAN the BIO LAB.

H.S. "SCIENCE FICTION CLUB" AND FEATURES
EDITOR of its OFF-CAMPUS "FANZINE"
ENTITLED ETAOIN SHRDLU.

In (elementary/secondary) school, I entered science fair(s).
_____*YES*_____. If yes, briefly describe the nature of the competition(s),
  (yes/no)
how old you were when it/they took place, and indicate how you fared
(first, second, no prize, etc.).

ELEM. School — N.Y. SCIENCE FAIR
3RD PRIZE
(I MADE A STROBOSCOPE
AND DEMONSTRATED ITS UTILITY)

1

In (elementary/secondary) school, I entered the following science
competitions (include local, national, and international contests)
_____✓_____. If yes, briefly describe the nature of the competition(s),
   (yes/no)
how old you were when it/they took place, and indicate how you fared
(first, second, no prize, etc.).

16 yrs old — WESTINGHOUSE SCIENCE
TALENT SEARCH. I WAS
one of 40 FINALISTS. [MY
project — 'The Substitution of SELENIUM
for SULFUR IN HYDROponic plant
Growth' RECIEVED NO ASSISTANCE, LAB space,
TEACHER TIME, etc. from my school.]

Would you be willing to share with our readers your thoughts about the
influence of science competitions on your development as a scientist?
Do you endorse the idea of holding competitions or not?

Such competitions REpresented
the only ENCOURAGEMENT I EVER
recieved from the ED-BIZ
ESTABLISHMENT to pursue
SCIENCE AS A CAREER.

You __MAY__ publish my responses with attribution.
   may/may not
Signature _____
   (date)
            GLASHOW
Thank you very much for taking the time to help us gather these data.

Oct 15 89   2